MW00844934

M

B

Steam Engine

Design & Mechanism

International Correspondence Schools

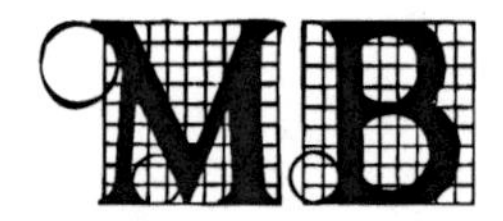

ISBN 1-60386-110-6

STEAM-ENGINE MECHANISM

ELEMENTS OF THE STEAM ENGINE

FUNDAMENTAL PARTS

1. The Four-Link Slider Crank.—The steam engine is a mechanism for transforming the energy of steam into work. In *Entropy and Steam*, it was shown that steam may do work by lifting weights against the pressure of the atmosphere. As it is not generally desired to do work in this manner, it is essential that some method of changing the to-and-fro motion of the piston into a continuous motion in one direction should be devised. The form of mechanism used for this purpose in most types of engines is shown in Fig. 1.

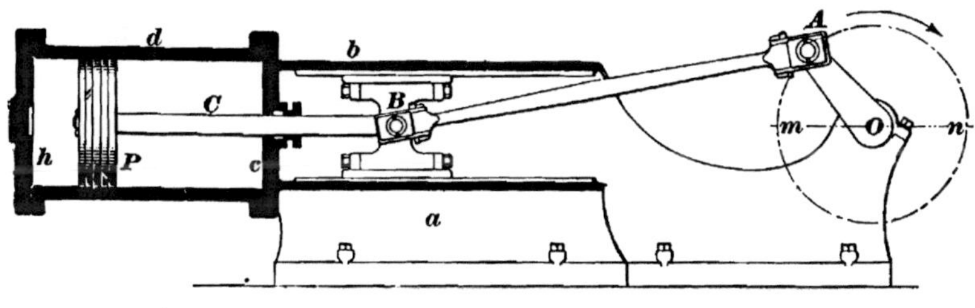

Fig. 1

It is composed of four parts or **links:** the link OA, called the **crank;** BA, the **connecting-rod;** CB, the **piston rod;** and the stationary link a, called the **frame.** The part shown at b is called the **guide,** and that at d, the **cylinder.** The cylinder, guide, and frame are rigidly connected to one another, and form the fourth or fixed link. These four links form what is known as the **four-link slider crank.**

2. The piston P moves first to one end and then to the other end of the cylinder d. The steam from the boiler enters one end—say, in this case, the end h—of the cylinder, and pushes the piston to the other end. By means of another mechanism called the **valve,** the steam is now admitted to the end c of the cylinder, while the end h is at the same time allowed to communicate with the atmosphere or with a condenser. The steam in h escapes, while that in c pushes the piston back again to its original position, whence the same operation is repeated.

Attached to the piston, and forming a part of it, is the piston rod CB; to the end of the piston rod is fastened, by a joint, one end of the connecting-rod BA. The other end of BA is joined to the crank OA; and the other end of OA terminates in a shaft O, which rotates in stationary bearings. It is evident that the end of BA, which is attached to CB, can move only in a straight line; and since the shaft O can rotate only in its bearings, the end of OA, which is attached to BA, can move only in a circle.

When the piston P is at one extreme end of the cylinder, say at h, the joint A is at the point m, and all three links, OA, BA, and CB, lie in a straight line. As the piston moves to the right, the link CB also moves to the right, while the joint A is constrained to move in the upper semicircle mn; when P reaches the other end of the cylinder, the joint A is at n, and again OA, BA, and CB are in a straight line. The piston now moves back to the end h of the cylinder, the joint A moving in the lower semicircle from n to m.

3. Those parts of the four-link slider crank that have a to-and-fro, or reciprocating, motion are called the **reciprocating parts.** They are the piston, piston rod, crosshead, and connecting-rod.

The end h of the cylinder is called the **head end,** and c the **crank end.** That is, the end of the cylinder farther from the crank is the head end, and the one nearer the crank is the crank end.

The distance passed through by the piston in moving from one end of the cylinder to the other while the crank is making half a revolution is called the **stroke;** the stroke is evidently equal to the diameter mn of the circle described by the end A of the crank.

The engine may run in the direction shown by the arrows in the figure, or it may run in the reverse direction. In the former case, it is said to **run over;** and in the latter case, to **run under.** In other words, an engine runs over when the crank passes through the upper half of its circle as the piston moves from the head end to the crank end of the cylinder; and the engine runs under when the crank passes through the lower half of its circle as the piston moves from the head end of the cylinder to the crank end.

The stroke from the head end to the crank end of the cylinder—that is, from left to right in the figure—is called the **forward stroke;** the one from crank end to head end, the **return stroke.**

The foregoing mechanism gives a continuous motion in one direction. A pulley is keyed to the shaft O, and the power is transmitted by belting to shafting, or directly to the machinery to be run.

THE PLAIN SLIDE-VALVE ENGINE

4. Types of Engines.—There are many types of steam engines, most of which are covered by the following classification:

1. According to the kind of service, as	Stationary Locomotive Marine
2. According to number and arrangement of cylinders, as	Simple Compound Triple expansion Quadruple expansion Duplex
3. According to the type of valve used, as	Plain slide valve Automatic cut-off Corliss

Any of these may be *horizontal* or *vertical*, *condensing* or *non-condensing*, *single-acting* or *double-acting*. All these types involve essentially the same principles, and therefore the description of a single type will be sufficient to give a general knowledge of these principles.

5. Parts of the Engine.—The various parts of a steam engine are shown, in their relations to one another, in Fig. 2, which represents a simple slide-valve engine. In this figure, *1* is the cylinder; *2* is the cylinder head, and the ends of the cylinder at *2* and *3* are known as the head and crank ends, respectively; *4* is the piston; *5* is the piston rod; *6* is the crosshead; *7* is the connecting-rod; *8* is the crosshead pin; *9* is the crankpin; *10* and *11* are crank-disks; *12* is the fly-wheel; *13* is the crank-shaft, or main shaft; *14* is one of the main-shaft bearings; *15* is a belt pulley; *16* is the eccentric; *17* is the eccentric strap; *18* is the eccentric rod; *19* is the valve stem; *20* is the slide valve; *21* is the steam chest; *22* and *23* are the steam ports; *24* is the exhaust port; *25* is the frame or bed of the engine; *26* and *27* are the upper crosshead guides; *28* is the exhaust pipe, which connects with the exhaust port; *29* is the steam pipe; *30* is the throttle valve; *31* is the governor, which controls the speed of the engine by regulating the pressure of the steam admitted to the steam chest through a valve in the casing *32*; *33* is the governor pulley, driven by a belt *34* from the main shaft.

In the upper portion of the figure is shown a section of the cylinder and steam chest removed so as to show the interior of the cylinder.

Among the details of the engine, *35* is the lagging of the cylinder; *36* is the piston-rod stuffingbox; *37* is the valve-stem stuffingbox; *38* is the valve-stem slide; *39* is an oil-hole cap; *40* and *41* are holes tapped into the cylinder for the purpose of attaching indicator piping; *42* is the steam-chest cover.

6. A section of a steam-engine cylinder is shown in Fig. 3. The working length l of the cylinder is slightly less than the distance between the cylinder heads, since a small

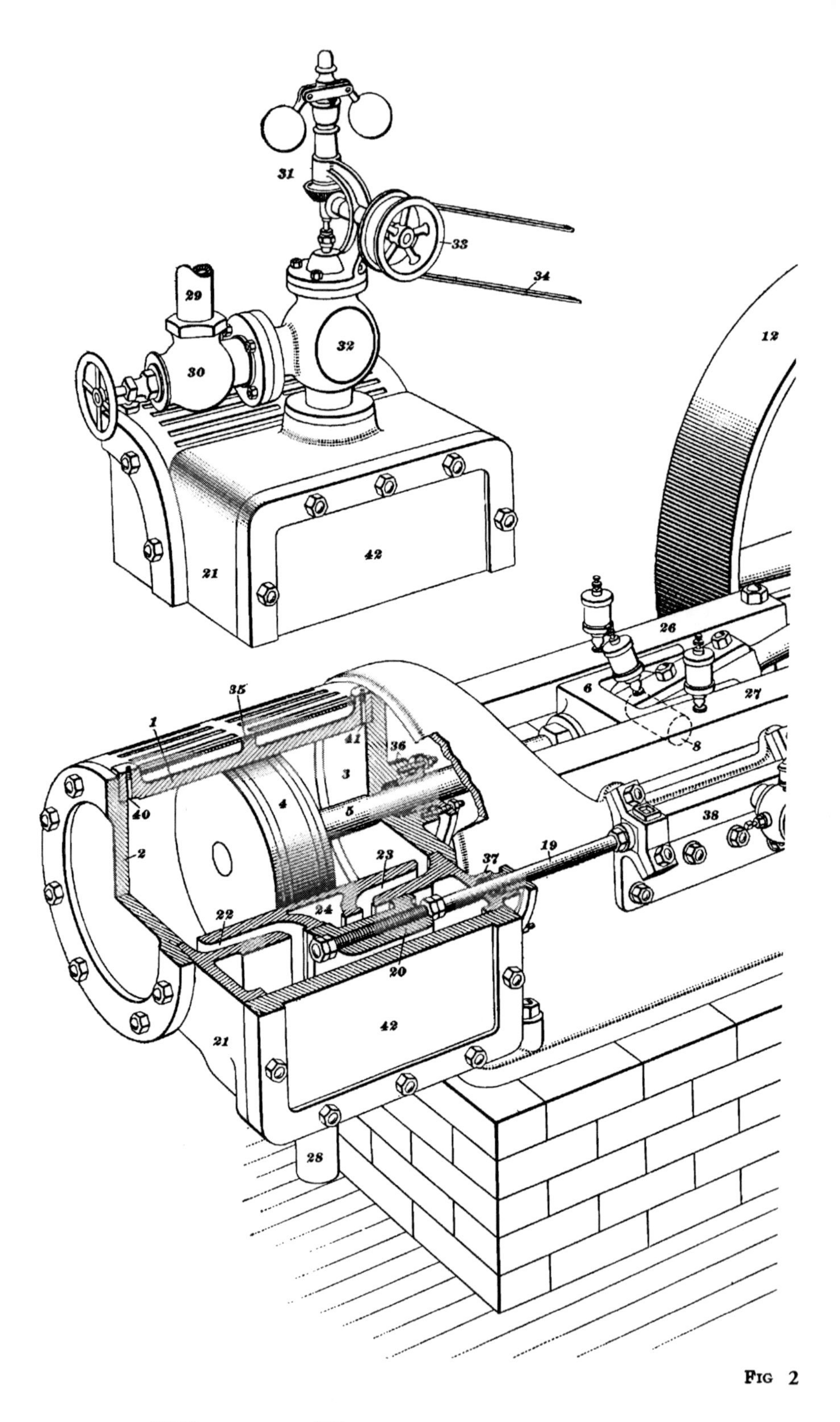
31
33
34
29
30
32
12
42
21
26
6
27
35
1
41
36
3
8
40
4
5
38
2
23
37
19
22
24
20
42
21
28

Fig 2

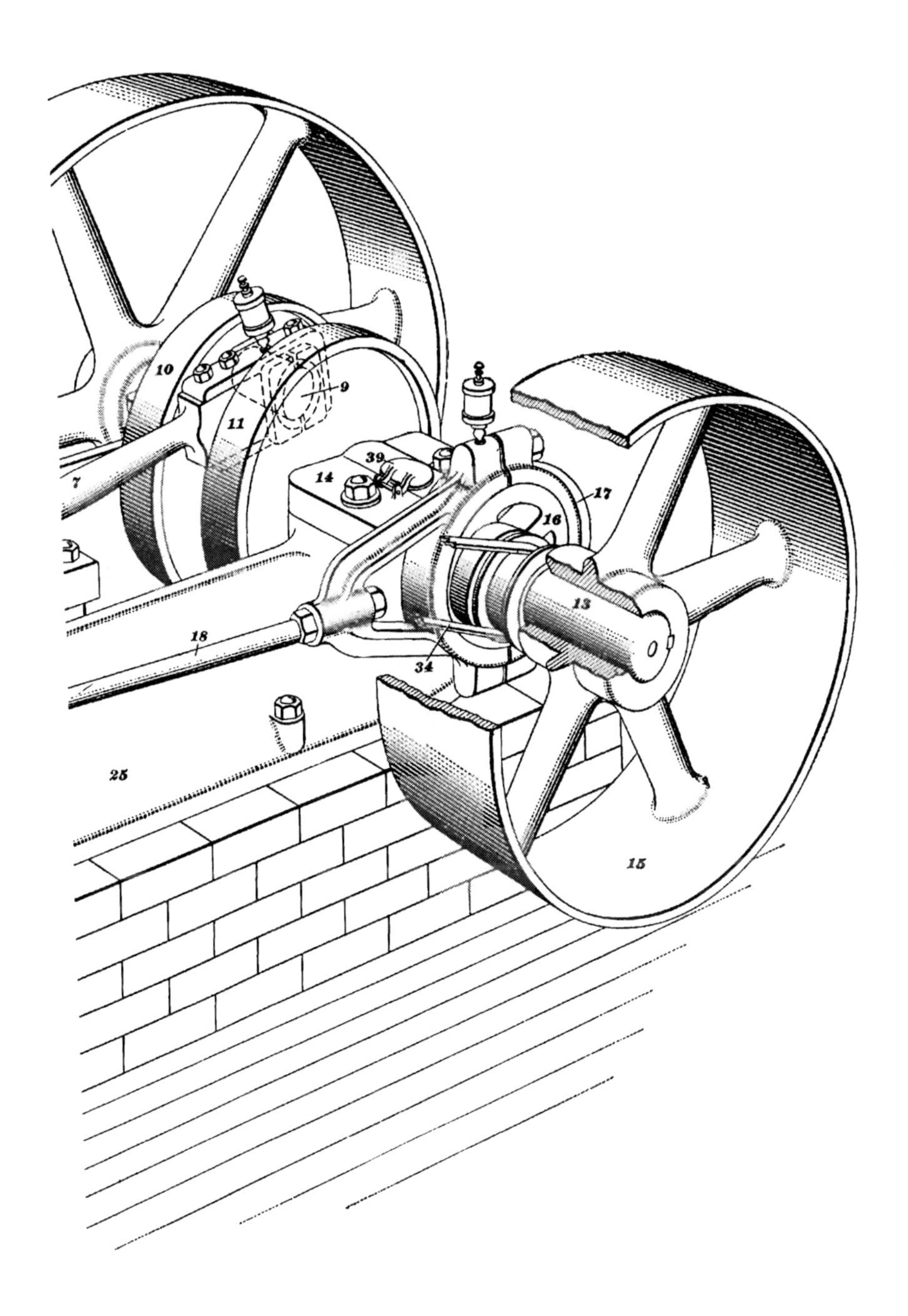

10
9
11
7
39
14
17
16
13
18
34
25
15

space must be left between the head and the piston when the latter is at the end of its stroke. The volume of this space, together with the volume of the steam port that leads to it, is called the **clearance.**

The **stroke** of the engine is the travel of the piston *p*; since the piston and crosshead are rigidly fastened to the same rod, the stroke must also be equal to the travel of the crosshead. It was shown in Fig. 1 that the stroke is also

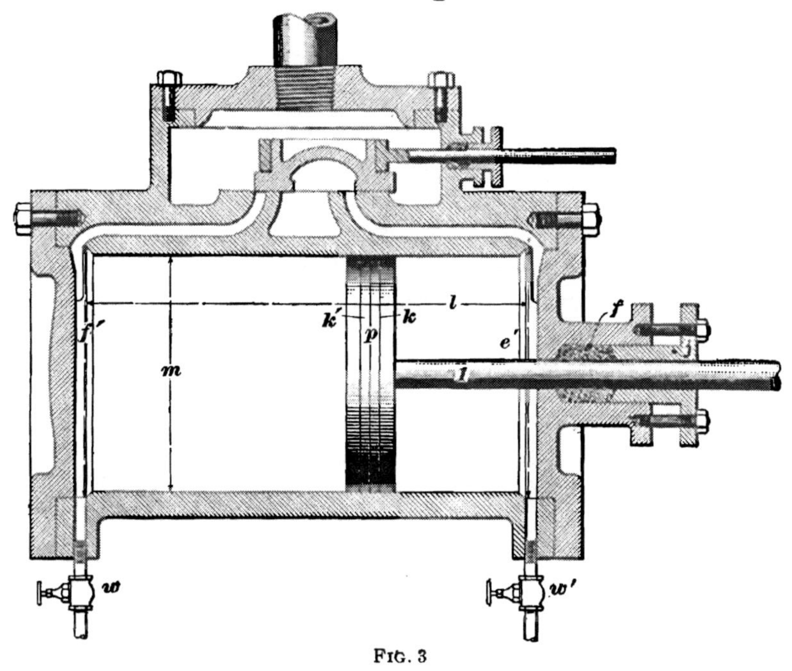

Fig. 3

equal to the diameter of the circle described by the crankpin, or what is the same thing, equal to twice the length of the crank, this length being measured from the center of the crankpin to the center of the crank-shaft. The diameter or **bore** of the cylinder is represented by *m*.

The size of an engine is generally expressed by giving the diameter of the cylinder and the stroke in inches. Thus, an engine having a cylinder diameter of 16 inches, and a stroke of 22 inches, is called a 16″ × 22″ engine. In stating the size of an engine in this way, the diameter of the cylinder must be given first and the stroke next.

At the ends e' and f', the cylinder is **counterbored;** that is, for a short distance the bore is greater than m. The piston projects partly into this counterbore at the end of each stroke. Were it not for the counterbore, the piston would not wear the cylinder walls their entire length, and shoulders would be formed at each end of the cylinder. When the wear of the joints in the connecting-rod is taken up, the length of the connecting-rod is changed, and the piston is moved slightly from its original position. In this case, a shoulder would cause an undesirable pounding of the piston.

Drain cocks w, w' are fitted in each end of the cylinder, through which any condensed steam may be discharged.

7. The piston fits loosely in the cylinder and has split rings k, k' inserted, which springs out so as to press against the wall of the cylinder and prevent leakage of steam between the wall of the cylinder and the piston. Pistons are usually supplied with a follower plate, which is bolted to the head end of the piston in order to hold the split rings k, k' in place. The piston rod is a round bar rigidly connected to both the piston p and the crosshead; f is a stuffingbox in which packing is placed, and is fitted with a gland j, which, when bolted down, compresses the packing around the piston rod and makes a steam-tight joint. This packing is often made in the form of split rings, which are so placed that the split of the first ring is covered by the solid part of the next ring. When repacking, care should be taken not to cause unnecessary friction by too much pressure from the gland. The crosshead, shown at *6*, Fig. 2, is given an easy sliding fit between the guide bars *26*, *27*, which are in line with the path of the piston rod, and, with the crosshead, relieve the piston rod of all bending strains.

The connecting-rod *7* forms the connecting link between the crosshead and the crank-disks. The joint between the crosshead *6* and the connecting-rod *7* is made by the crosshead pin *8*; and the joint between the connecting-rod and the crank-disks is made by the crankpin *9*. Connecting-rods are usually made from four to six times the length of the crank, or from *4 to 6 cranks* in length, as it is called.

8. The eccentric, which imparts motion to the slide valve, is shown in Fig. 4. It consists of a circular disk of

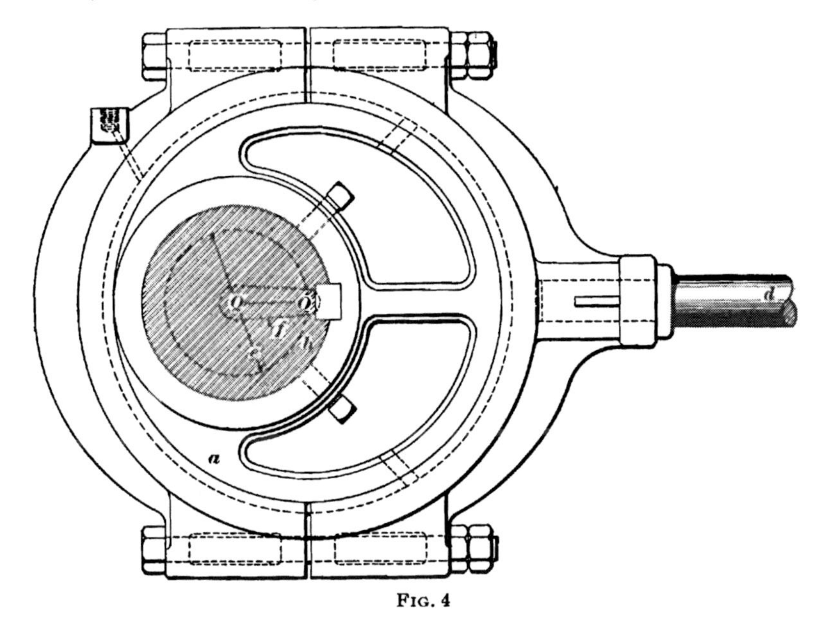

Fig. 4

iron *a*, which is keyed or fastened by setscrews to the shaft and revolves with it. The center *O* of the eccentric is at some distance from the center *Q* of the shaft, so that, as the shaft turns, the center *O* of the eccentric describes the circle *b* whose diameter is *e*. The **eccentric strap** *c*, which surrounds the eccentric, is fastened to the eccentric rod *d*. Hence, for each half revolution of the shaft, the eccentric rod is moved horizontally a distance equal to the diameter *e*. This distance is called the **throw** of the eccentric. The distance *O Q* from the center of the eccentric to the center of the shaft is called the radius of the eccentric, or the **eccentricity.** Evidently the throw is twice the eccentricity. Some engineers consider the throw as being equal to the radius *O Q*, but throughout the succeeding pages the definition as given above will be observed.

The eccentric is equivalent to a crank whose length is equal to the eccentricity. Thus, if the end of the eccentric-rod *d* were attached at *O* to the crank *f*, shown by dotted lines, the motion imparted by the crank would be precisely like that given by the eccentric.

In practice, the diameter of the shaft generally exceeds the diameter *e* of the circle described by the eccentric. In plain slide-valve engines, the eccentric is usually keyed to the shaft after being properly adjusted. The connection between the eccentric rod and the valve stem is accomplished in a variety of ways. In Fig. 2 a slide *38* is used to support the joint between the eccentric rod *18* and the valve stem *19*. The latter must be supported in some such manner to prevent it from binding in its stuffingbox.

THE D SLIDE-VALVE AND STEAM DISTRIBUTION

9. The Valve and Valve Seat.—Of the different kinds of valves used to distribute the steam in the engine cylinder, the **D** slide valve is the most common. A section of such a valve is shown in Fig. 5; *p* and *p* are the **steam ports,**

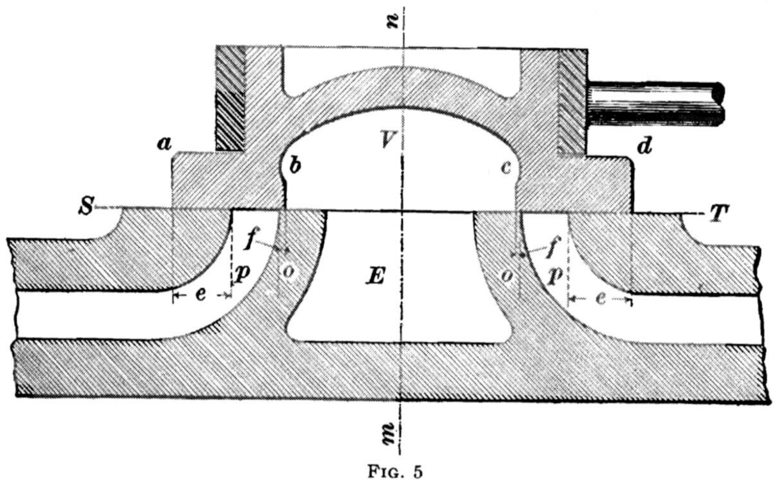

Fig. 5

o and *o* the **bridges,** *E* the **exhaust port,** and *S T* the **valve seat.** The valve flanges *a b* and *c d* extend beyond the edges of the steam ports in both directions. The distance by which the edge of the flange extends beyond the outer edge of the steam port, when the valve is in its central position, is called the **outside lap.** In the figure, *e*, *e* represent the outside laps for the two ends of the cylinder. The distance by which the edge of the flange extends beyond the inner edge of the steam port, when the valve is in its central position, is called the **inside lap** and it is represented in the figure by *f*, *f* for the two ends of the cylinder.

The valve is here shown in mid-position; that is, the center line *n* of the valve coincides with the center line *m* of the exhaust port. As the motion of the valve is caused by the eccentric, the valve is in mid-position when the radius *Q O* of the eccentric, Fig. 4, is in a vertical position. When *Q O* lies horizontally on the right side of *Q*, the valve is in its position nearest the head end of the steam chest, and when *Q O* lies horizontally on the left side of *Q*, the valve is at the end of its stroke in the crank end of the steam chest. A valve is said to have **lead** when the steam port is opened slightly before the piston reaches the end of its stroke.

10. Relative Position of Valve and Piston.—In order to show how the steam is distributed in the cylinder by means of the valve and eccentric, a series of skeleton diagrams have been drawn showing the relative positions of the valve and piston for different points of a double stroke. Fig. 6 shows five diagrams representing a **D** slide valve without lap or lead. *O a* represents the crank; *O b* the eccentric; *a c* the connecting-rod; and *b d* the eccentric rod. It should be remarked that the sizes of some of the parts have been greatly exaggerated, particularly the radius of the eccentric circle and the amount of clearance. Diagram *A* represents the piston as just on the point of beginning the forward stroke. The valve is moving in the direction of the arrow and the outer edge is just about to admit steam to the left-hand port. As will be seen, the valve is in its central position and consequently the line joining the center of the shaft and the center of the eccentric, which represents the *eccentric radius* is vertical. All the parts are moving in the direction of the arrows.

Diagram B shows the position of the parts when the crank

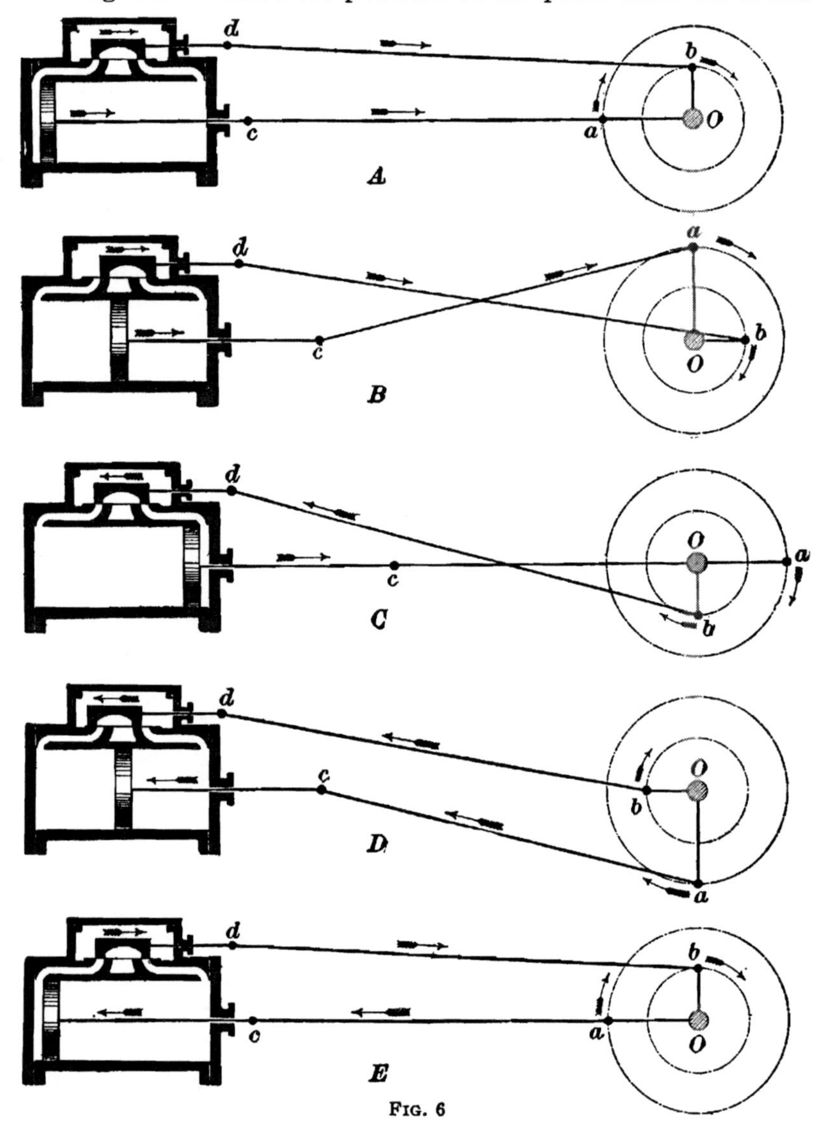

FIG. 6

has moved through 90° from its position in A. The piston is at the middle of its stroke, or very nearly there. It would be exactly at the middle of its stroke but for the fact that the connecting-rod makes an angle with the horizontal. The angularity of the connecting-rod will be treated of later; for the present it will be assumed that it has no effect on the position of the piston. The valve has reached the extreme limit of its travel to the right, and the eccentric radius *O b* is horizontal. The left steam port is fully opened for the live steam, and the right steam port is fully opened for exhaust.

Another crank-movement of 90° places the different parts as shown in diagram *C*. The piston has reached the end of its forward stroke; the valve, which is in its central position, moving toward the left, has just closed the left steam port and right exhaust port, and is just about to open the right port for the admission of live steam, and the left port for the release of exhaust steam. The piston has now traveled one full stroke.

Diagram *D* shows the piston in its central position on the return stroke. The crank is in the position *O a*; the eccentric is horizontal, as represented by *O b*, and the valve is at the farthest point of its travel to the left, the right port being fully open for live steam and the left port fully open for exhaust.

In diagram *E*, the piston has reached the extreme point of the return stroke, the piston-rod, connecting-rod, and crank being all in one straight line; this also occurs in diagrams *A* (which is the same as *E*) and *C*. The valve has been moving to the right, and is now in its central position, just on the point of admitting steam to the left port.

11. These diagrams show conclusively that, with no lap or lead, the steam is admitted to the cylinder for the full stroke of the engine, consequently there can be no cut-off, and, therefore, no expansion of steam.

The following conclusion is now evident: *With an ordinary* **D** *slide valve, operated by one eccentric, there can be no cut-off until the end of the stroke, and, therefore, no expansion of steam, unless the valve has outside lap.*

12. Distribution of Steam.—The effect of lap on the movement of the valve relatively to the piston, and also on the movement of the eccentric and crank, is shown **in Figs.** 7 to 14. In these figures, the valve has both outside and inside lap, but no lead. These diagrams have been distorted, as was done in Fig. 6, in order that the eccentric radius might be long enough to show the relative positions of the moving parts. In Figs. 7 to 14, the eccentric radius is three times as long as it should be for the amount of valve movement shown by the figure. The diameter of the crank-circle is also a little greater than the stroke of the piston for the same reason. In order to show the distribution of steam by the valve, a diagram has been drawn above and below each cylinder, those above being marked *M*, and those below, *N*. These diagrams are supposed to be drawn in the following manner: Imagine it to be possible to connect two small pipes to the piston, one on each side. Suppose that each pipe has a steam-tight piston working in it, the lower sides of the pistons being subjected to the steam pressure in the cylinder, and the upper sides to the atmospheric pressure. Suppose, further, that there is a coiled spring on top of the piston; that a piston rod passes through the center of the spring; and that a pencil is attached to the end of the piston rod. If a pressure of 10 pounds is required to compress the spring 1 inch, it is evident that for every 10 pounds pressure in the cylinder the pencil will move upwards 1 inch, and that if it touched a sheet of paper it would mark a line on that paper. It will now be presumed that an arrangement like that just described is attached to the steam-engine piston, and that the pencil touches a sheet of paper, which is held stationary. Then, when the steam piston moves ahead, the pencils will make straight lines at heights corresponding to the steam pressure on the under sides of the little pistons, except when the pressure of the steam in the cylinder varies, in which case the pencil will move up or down, according as the pressure increases or diminishes.

Having made these suppositions clear, let QX, Figs. 7 to 14, represent the line that the pencil would trace if there were a perfect vacuum in the cylinder; that is, QX is the line of no pressure; also, let AB represent the line that the pencil would trace if the pressure in the cylinder were just equal to that of the atmosphere, and *Q Y* were the line of no volume. Then the point *Q* represents no volume and no pressure. Finally, let *1 D* represent the volume of the clearance.

13. In Fig. 7, the piston is represented as just beginning

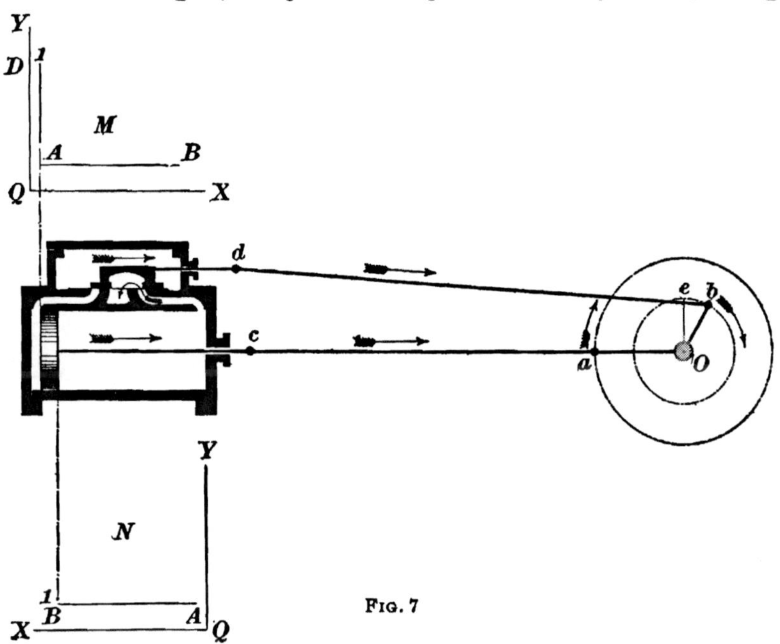

FIG. 7

the forward stroke, and the valve as just commencing to open the left steam port, both moving in the same direction, as shown by the arrows. If the valve had no outside lap, the position of the eccentric center would be at *e*; but on account of the lap, the valve has moved ahead of its central position in order to bring its edge to the edge of the port. To accomplish this, the eccentric center has been moved from *e* to *b*, *O b* being the position of the eccentric radius. The angle *b O e* that the eccentric radius makes with the position it would be in if there were no lap or lead, is called the **angle of advance.** In other words, the angle of advance is the angle between the position of the eccentric radius when the valve is in mid-position, and its position when the piston is at the end of the stroke. It is equal to the angle due to the lap plus the angle due to the lead; it is frequently called **angular advance.**

Assume that the piston and valve have moved a very small distance, just sufficient to admit steam to fill the clearance space on the left of the piston, so that the steam acts on the piston at full boiler pressure. If the length of the line *A 1* represents the boiler pressure (gauge), the pencil that registers the pressure on the left side of the piston will be at *1*.

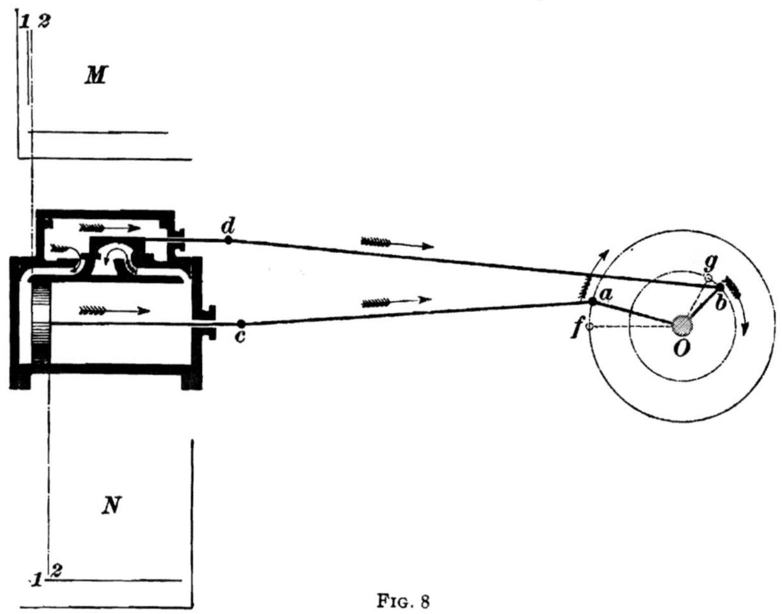

FIG. 8

The steam on the right side of the piston is exhausting into the atmosphere through the exhaust port, as shown by the arrow. As the size of the exhaust port is limited by practical considerations, the exhaust is not perfectly free, and there is consequently a pressure on the exhaust side of the piston; this is termed **back pressure.** Therefore, in the diagram *N*, let *1* be the position of the second pencil; then *1 B* is the back pressure.

14. Fig. 8 shows the position of the piston and valve when the exhaust is fully open. The crank has moved from the position *O f*, shown by the dotted line, to *O a*, and the eccentric center from *g* to *b*. Steam is entering the head end of the cylinder and exhausting at the crank end. The pencils have moved from *1* to *2* on both diagrams *M* and *N*.

15. In Fig. 9, the piston has advanced far enough to enable the valve to reach the end of its stroke and open the port to its full width. The crank and eccentric have moved to the positions *Oa* and *Ob*, the dotted lines showing their

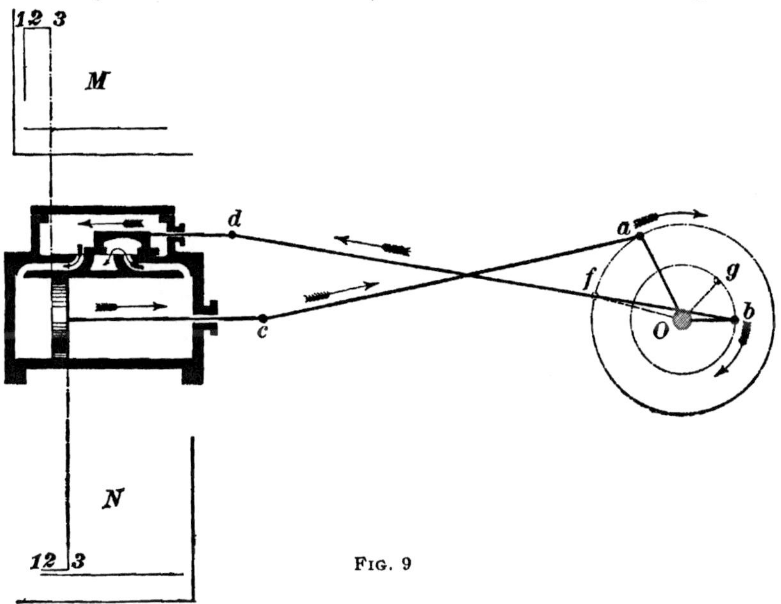

Fig. 9

last, Fig. 8, position. The eccentric radius is horizontal and any further movement of the crank will cause the eccentric to travel in the lower half of its circle and make the valve move back. In diagrams *M* and *N*, the pencil has traced the lines *2–3*.

16. Fig. 10 shows the piston still further advanced on its stroke, and the valve as having its inside edge in line with the outside edge of the right steam port. The left end of the valve has partially closed the steam port. The amount of advancement of the crank and eccentric from their last positions is shown by their distances from the dotted lines. The pencils have traced the lines *3–4* on the diagrams *M* and *N* during this movement of the piston from the last position.

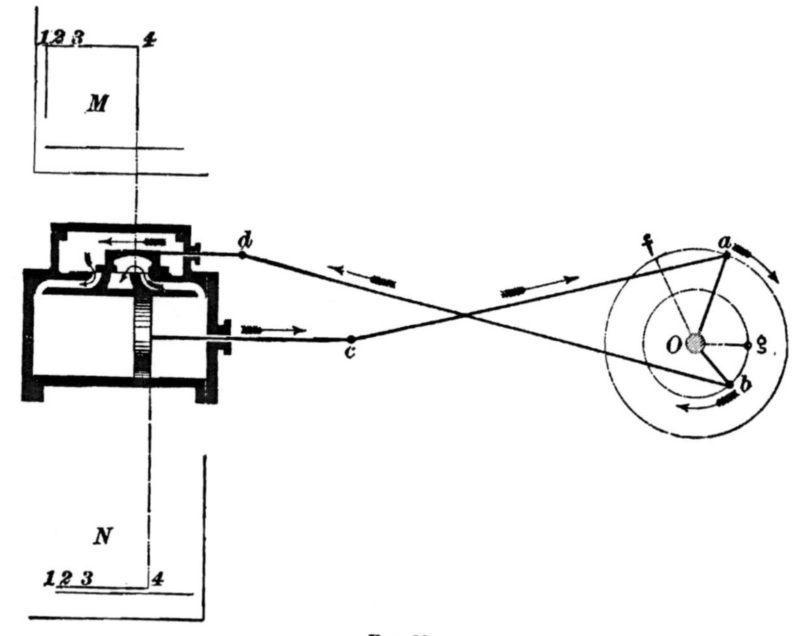

FIG. 10

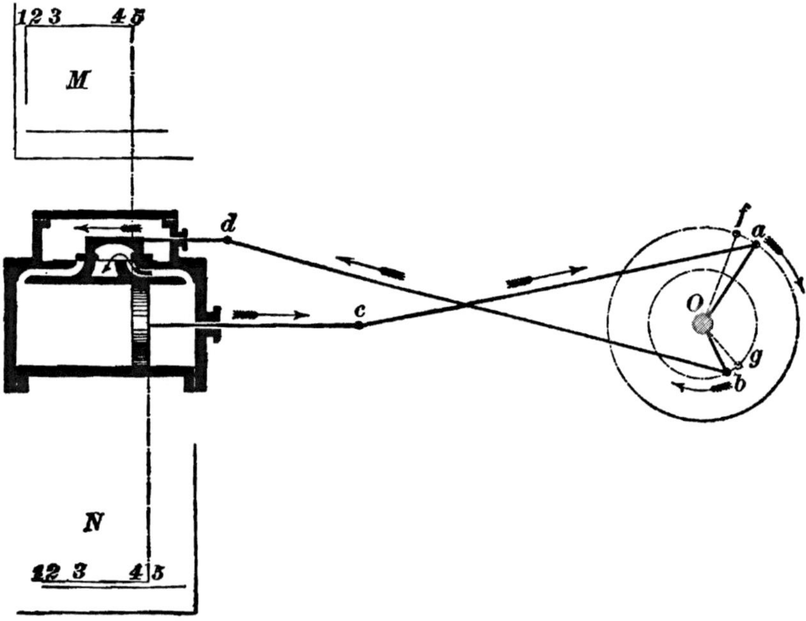

FIG. 11

Fig. 11 marks one of the most important points of the stroke. Here the valve has closed the steam port, that is, cut off the steam, and from here to the end of the stroke the steam in the cylinder expands. The exhaust is now partially closed. The crank and eccentric have moved through the angles indicated. During this movement the pencils have traced the lines *4–5*.

17. Fig. 12 shows another very important valve position.

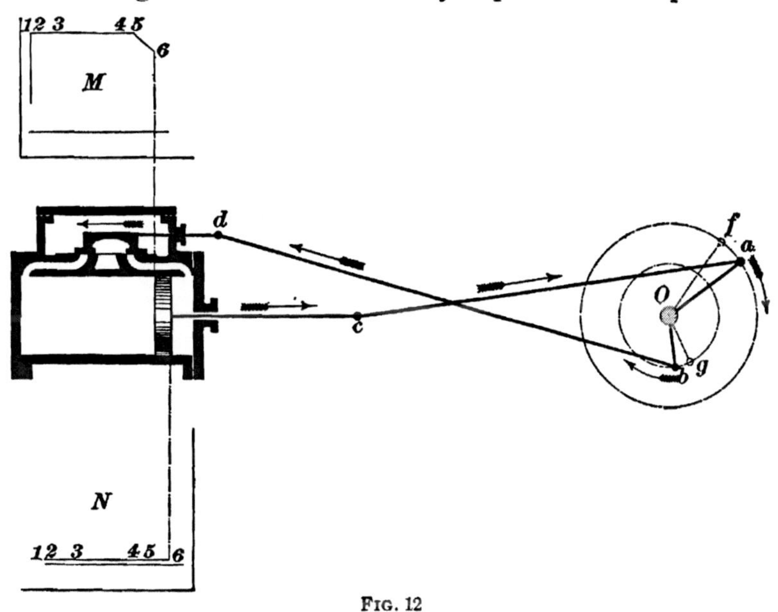

Fig. 12

Here the inside edge of the valve closes the exhaust, and from now on to the end of the stroke the steam in front of the piston is compressed. In the diagrams *M* and *N*, the lines *5–6* are traced by the pencils. The line *5–6* on the diagram *M* is an expansion line, the pressure falling as the piston moves ahead.

18. In Fig. 13, the piston has advanced far enough to cause the left inside edge of the valve to be in line with the inside edge of the left port. The slightest movement of the valve to the left will open the left port to exhaust. Expansion really ends here, although, on account of the limitation in the size of the ports, there will still be a slight further expansion owing to the inability of the steam to escape instantly. During this last movement of the piston, the pencils trace the lines *6–7* on the diagrams *M* and *N*. On the diagram *M*, the line *6–7* is a continuation of the expansion

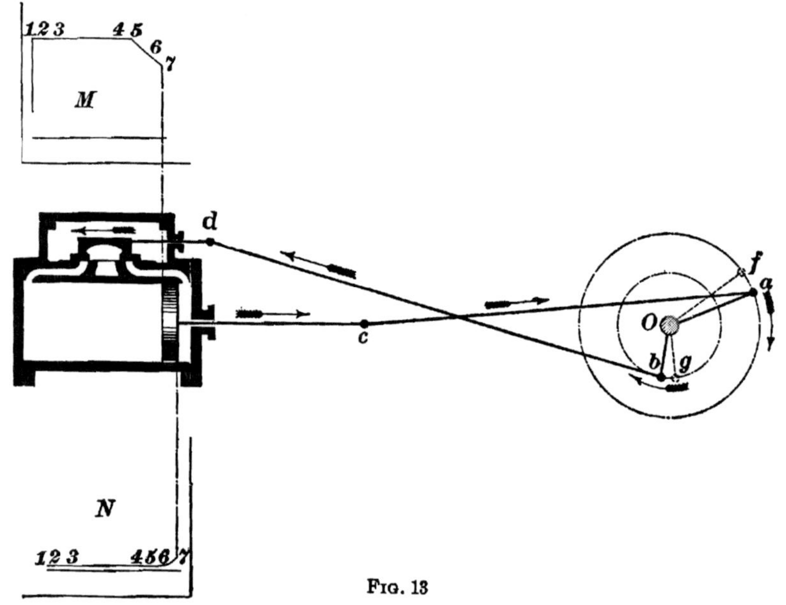

FIG. 13

line *5–6*; while on the diagram *N* it shows part of the compression line, the pressure rapidly increasing as the piston nears the end of the stroke.

19. In Fig. 14, the piston has reached the end of its forward stroke and is about to begin the return stroke. The right outside edge of the valve is in line with the outside edge of the right port. The steam is exhausting from the head end of the cylinder, as shown by the arrows. The crank and eccentric are both diametrically opposite their positions in Fig. 7. In the diagrams *M* and *N*, the pencils have traced the lines *7–8*. *M* shows that the pressure has fallen very rapidly from *7* to *8*; while in *N*, it has risen from *7* to *8*. The very slightest movement of the piston to the left will admit steam to the crank end of the cylinder and cause the pencil to rise to the point *1′*.

20. During the return stroke, the actions of the steam just described will be repeated, the pencils tracing the dotted lines on the diagrams *M* and *N*, Fig. 14, the exhaust going through the left port and the steam through the right port. As the process is so nearly like the preceding, the diagrams have not been drawn, but the valve should be followed

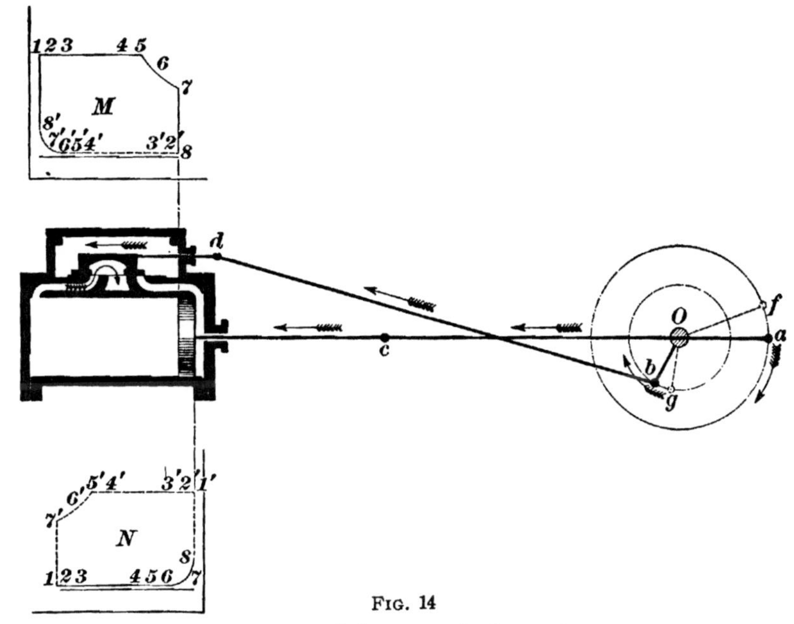

Fig. 14

through the different positions and the effects noted on the diagrams; to assist in this, the corresponding points have been numbered as in the foregoing figures.

21. Effects of Lap.—The study of Figs. 7 to 14 should show the effects caused by varying the lap. Thus, in Fig. 11, it is evident that if the outside lap were less, the valve would not close the left port when its center was in the position shown; consequently, the piston would have to move farther ahead before the valve could move back far enough to close the port. This, of course, would make the cut-off take place later in the stroke and shorten the expansion. It is likewise evident that if the valve had more lap, this extra lap would extend beyond the port when the center of the valve was in the position shown. Therefore, the valve would cut off earlier in the stroke and the expansion would be lengthened. Hence, *increasing the outside lap means an earlier cut-off and an increased expansion, while decreasing the outside lap means a later cut-off and a diminished expansion.*

Considering the inside lap, it is evident, from Fig. 12, that if the inside lap had been less, the exhaust would not have closed so soon, and consequently the compression would have begun later; had the inside lap been greater, the compression would have begun earlier. Fig. 13 shows that with a diminished inside lap, the opening of the exhaust, usually termed **release,** would take place earlier; while with an increased inside lap, the release would have taken place later in the stroke. Hence, *increasing the inside lap increases compression and delays release, while diminishing the inside lap decreases compression and hastens release.*

22. Lead.—A valve is said to have **lead** when it com-

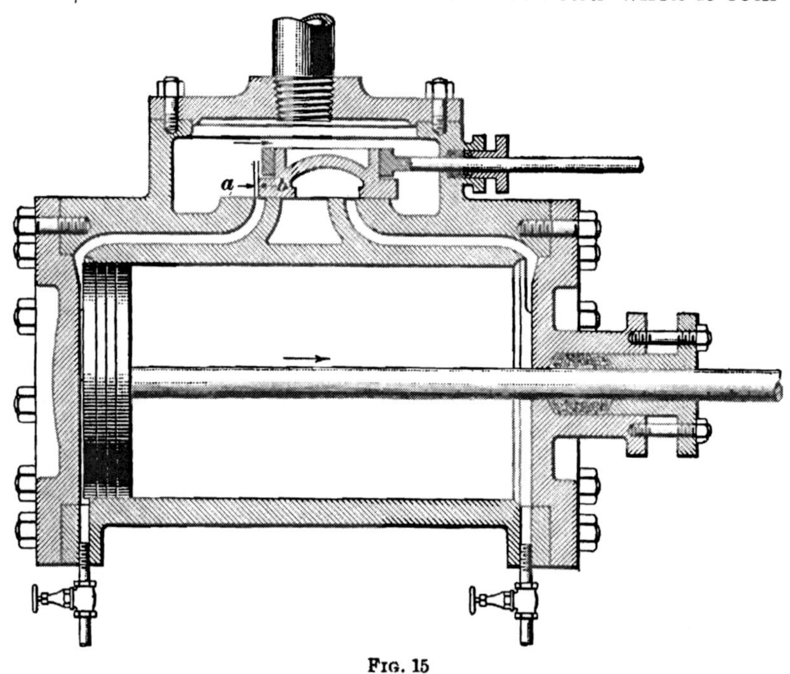

FIG. 15

mences to open the steam port just before the piston reaches the end of the stroke. The lead is the distance between the edge of the valve and the edge of the port from which the valve is traveling when the piston is at the end of its stroke. In Fig. 15, the distance ab is the lead. Lead is given to a valve in order to have the clearance space filled with steam at boiler pressure when the piston begins its stroke. The effect of lead on the angular advance of the eccentric is evidently the same as an increase of lap; that is, it increases the angular advance. Its effect on the distribution of steam will be discussed further on.

23. Positions of Eccentric for Opposite Directions of Rotation.—In the preceding discussion of steam distribution, it has been assumed that the engine runs over. When the engine runs under, the steam distribution and the piston and valve movements will be precisely the same as before, but the position of the center of the eccentric relative to the crankpin will be changed. To determine the position of the eccentric in this case, draw the horizontal diameter ae of the crank-circle, as shown in Fig. 16. Ob'

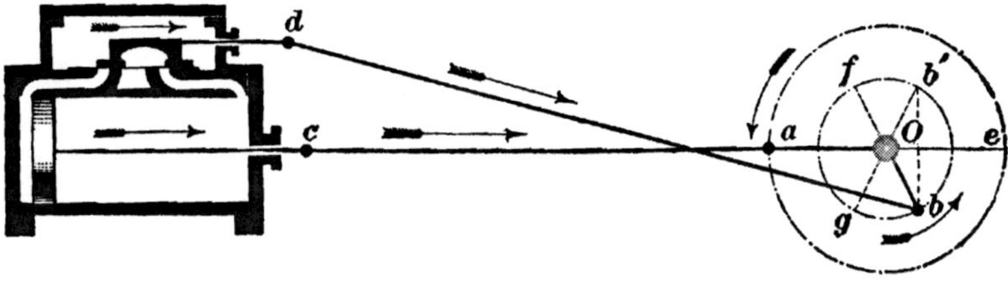

Fig. 16

represents the position of the eccentric radius when the piston is just beginning the forward stroke and the engine runs over. Draw $b'b$ perpendicular to ae and through the point b'; it intersects the valve circle in b, and b is the position of the center of the eccentric when the engine runs under and is about to begin the forward stroke. It is easy to see that this is so, for the valve and piston must both have a forward movement at this point, whether the engine runs over or under. If the eccentric radius were placed so as to occupy the position Of, the forward movement of the piston and downward movement of the crank would cause the valve to move to the left, closing instead of opening the port. It cannot be in the position *Og*, for that would throw the valve too far back. *Ob* is the only position in which the eccentric radius can be placed to give the valve the same movement when the engine runs under that it would have if placed in the position *Ob'* and the engine ran over. In both cases, the valve has the same forward movement, while the center of the eccentric is passing from *b* or *b'* to the horizontal position *Oe*.

24. Rocker-Arms.—It frequently happens that the eccentric cannot be so located on the shaft, that the eccentric rod and valve stem shall be in the same straight line. It can never be done when the valve is on top of the cylinder without inclining the valve seat, now very seldom done; and, with the valve on the side of the cylinder, other considerations, such as the location of the flywheel, may interfere. In such cases as this, a lever, or **rocker-arm,** may be employed.

There are two kinds of rocker-arms—direct and reversing. A *direct rocker-arm* is one in which the points of attachment of the valve stem and eccentric rod lie on the same side of the fulcrum of the rocker-arm, in consequence of which the direction of motion of the valve is always the same as that of the eccentric. A *reversing rocker-arm* is one having its fulcrum between the points of attachment of the valve stem and eccentric rod. With a rocker-arm of this class, the eccentric and the valve always move in opposite directions.

There are four conditions that may arise when using a rocker-arm: (1) The travel of the valve, the throw of the eccentric, and the direction of motion of the valve and eccentric may be the same as before. (2) The direction of motion of the valve and eccentric may be the same as before, but the travel of the valve may be greater than the throw of the eccentric. (3) The travel of the valve and the throw of the eccentric may be the same as before, but the eccentric may move in the opposite direction. (4) The travel of the valve may be greater than the throw of the eccentric, and the direction of motion may be opposite that in (1) and (2).

25. Sometimes the valve travel is such that if the eccentric were made to have the same throw it would be inconveniently large. In such a case the valve and its seat may be raised, the valve stem connecting to the rocker-arm at a higher point, as illustrated in Fig. 17. The direction of

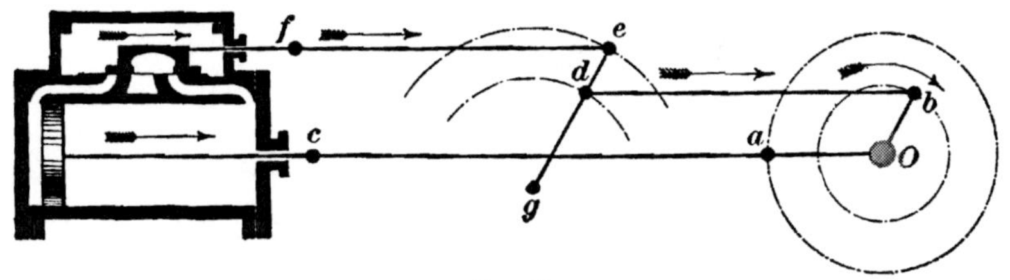

FIG. 17

motion of valve and eccentric is the same as in Figs. 7 to 14, but the throw of the eccentric is less than the travel of the valve by the ratio $gd : ge$; that is, if the valve travel is 4 inches, $gd = 12$ inches and $ge = 15$ inches, the throw of the eccentric will be $4'' \times \frac{12}{15} = 3.2$ inches. If the engine

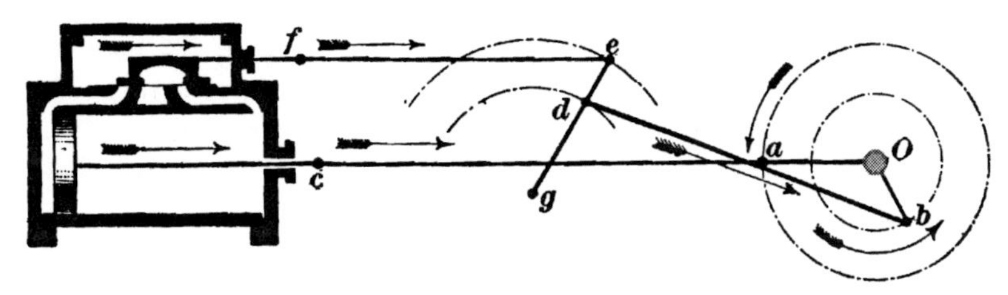

FIG. 18

runs under, the position of the center of the eccentric will be as shown in Fig. 18, and may be found by the same method as that given for finding it in the case shown in Fig. 16.

26. In the cases just described, the direction of motion of the valve and of the eccentric has remained the same as if there had been no rocker-arm, and both points of connection d and e, of the valve stem and eccentric rod, to the rocker-arm are on the same side of the pivot g. Suppose that the valve had been placed on the top of the cylinder, and it had been found more convenient to place the pivot of the rocker between the connections of the rocker-arm to the valve stem and eccentric rod, as shown in Fig. 19; then, when d moves to the right along the dotted arc whose center is at g, e moves to the left. Consequently, if the eccentric center were in the position $O\,b'$, and the engine were running in the direction of the arrow, the valve would move backwards instead of ahead. To overcome this difficulty, the

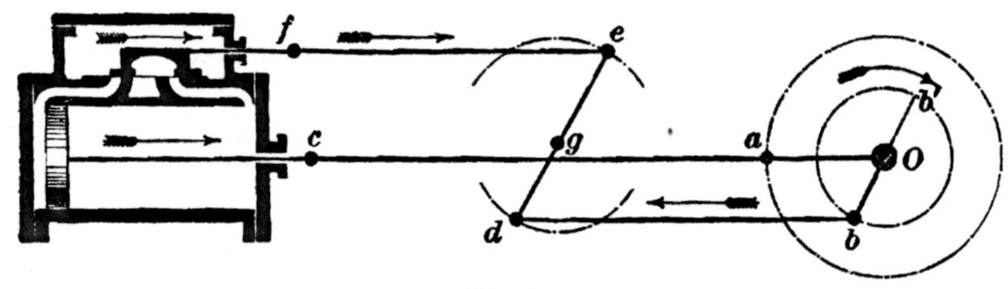

FIG. 19

eccentric is shifted around the shaft 180° to the position $O\,b$; then a movement of b in the direction of the arrow will throw d to the left and e to the right.

If $g\,d$ and $g\,e$ are equal, the valve travel and the throw of the eccentric will be equal, fulfilling condition 3. If $g\,d$ is less than $g\,e$, the throw of the eccentric will be less than the valve travel by the ratio $g\,d : g\,e$. For example, suppose that $g\,e$ = 20 inches and $g\,d$ = 15 inches and the valve travel is 5 inches; then the throw of the eccentric will be $5'' \times \frac{15}{20} = 3\frac{3}{4}$ inches.

27. Fig. 20 shows the position of the eccentric center when the engine runs under, and the rocker-arm is of the same design as the one shown in Fig. 19. If there were no

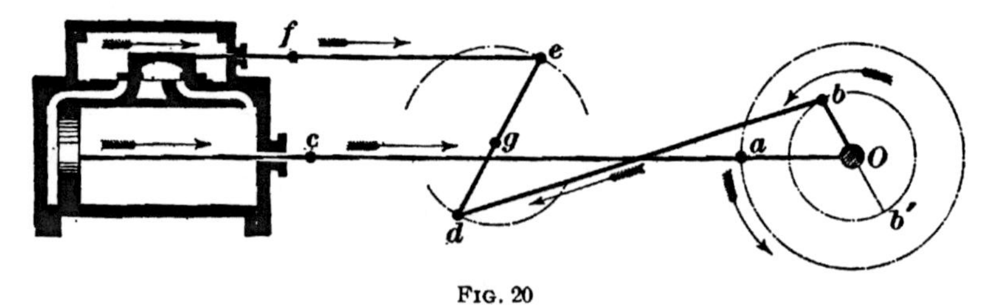

FIG. 20

rocker-arm, the eccentric center would be at b', as explained in Fig. 16, but, since the rocker-arm changes the direction of motion in this case, the eccentric is turned around 180°, to a point diametrically opposite.

The following rule may be applied to any engines whose valves cut off by their outside edges, as has been done in all the previous cases:

Rule.—*Place the crank in the position O a, and the eccentric in the position O b, as shown in Fig. 7, e O b being the angle of advance. If the engine runs over and the rocker-arm does not reverse the direction of motion of the eccentric, the eccentric is now correctly set. If the engine runs under, the eccentric should be placed in the position shown in Fig. 16, according to the rule given in connection with that figure. If the engine has a rocker-arm whose pivot lies between the point of connection with the valve stem and eccentric rod, and the engine runs over, place the eccentric center diametrically opposite the position shown in Fig. 7. If the engine runs under, and the pivot of the rocker-arm lies between the two points of connection, place the eccentric center diametrically opposite the position shown in Fig. 16.*

The following conveniently summarizes the instructions contained in the previous rule:

Direction of Running	Kind of Rocker-Arm	Angle Between Crank and Eccentric	Position of Eccentric Relative to Crank
Over	Direct	90° + angular advance	Ahead of crank
Over	Reversing	90° − angular advance	Behind crank
Under	Direct	90° + angular advance	Ahead of crank
Under	Reversing	90° − angular advance	Behind crank

28. Dead Centers.—When the piston has reached the end of either stroke, the piston rod, connecting-rod, and crank are all in one straight line, and the entire steam pressure on the piston is transmitted directly to the shaft and bearings, none of it being used to turn the crank. When the crank occupies this position, it is said to be on its **dead center.** This position is shown in Fig. 16. It is evident that there is no turning force on the crank due to the steam pressure when the reciprocating parts are in the position shown. There are two dead-center positions *Oa* and *Oe*, diametrically opposite each other, corresponding to the two extreme positions of the piston. When the crank occupies the position Oa, it is said to be on its **interior** dead center; and when it occupies the position Oe, it is on its **exterior** dead center. That is, when the crank is in line with the piston rod and connecting-rod, and lies on the side of the shaft toward the cylinder, it is said to be on its interior dead center; when on the opposite side of the shaft, it is said to be on its exterior dead center.

29. Clearance.—The term *clearance* is used in two senses in connection with the steam engine. It may be the distance between the piston and the cylinder head when the piston is at the end of its stroke, or it may represent the volume between the piston and the valve when the engine is on dead center. To avoid confusion, the former is called **piston clearance,** and the latter is termed simply **clearance.** Piston clearance is always a measurement, expressed in parts of an inch. Clearance, however, is a volume. Hereafter, then, clearance will be used to represent the volume of the clearance space. Wherever piston clearance is meant, it will be so stated.

When the crank is on a dead center and the piston at the end of its stroke, there is always a space between the piston and the cylinder head. The volume of this space plus the volume of the one steam port leading into it is called the clearance. Thus, in Fig. 15, the piston is at the end of its return stroke, and the clearance is the volume of the space between the piston and the left cylinder head, plus the volume of the left steam port. In other words, the clearance may be defined as the volume of steam between the valve and the piston, when the latter is at the end of its stroke. The clearance of an engine may be found by putting the engine on a dead center and pouring in water until the space between the piston and the cylinder head, and the steam port leading into it, is filled. The volume of the water poured in is the clearance.

The clearance may be expressed in cubic feet or cubic inches, but it is more convenient to express it as a percentage of the volume swept through by the piston. For example, suppose that the clearance volume of a 12″ × 18″ engine is found to be 128 cubic inches. The volume swept through by the piston per stroke is $12^2 \times .7854 \times 18 = 2{,}035.8$ cubic inches. Then, the clearance is $\frac{128}{2{,}035.8} = .063 = 6.3$ per cent. The clearance may be as low as ½ per cent. in Corliss engines, and as high as 14 per cent. in high-speed engines.

30. Theoretically, there should be no clearance, since the steam that fills the clearance space does no work except during expansion; it is exhausted from the cylinder during the return stroke, and represents so much dead loss. This is remedied, to some extent, by compression. If the compression were carried up to the boiler pressure, there would be very little, if any, loss, since it would then fill the entire clearance space at boiler pressure, and the amount of fresh steam needed would be the volume displaced by the piston up to the point of cut-off, the same as if there were no clearance. In practice, however, the compression is only made sufficiently great to cushion the reciprocating parts and bring them to rest quietly.

It is not practicable to build an engine without any clearance, owing to the formation of water in the cylinder due to the condensation of steam, particularly when starting the engine. As water is practically incompressible, some part of the engine would be broken when the piston reached the end of its stroke, if there were no clearance space for the water to collect in; usually, the cylinder heads would be knocked off. Automatic cut-off high-speed engines of the best design, with shaft governors, usually compress to about half the boiler pressure, and have a clearance of from 7 per cent. to 14 per cent.

Corliss engines require but very little compression, owing to their low rotative speeds; they also have very little clearance, since the ports are short and direct.

31. Real and Apparent Cut-Off and Ratio of Expansion.—The **apparent cut-off** is the ratio between the portion of the stroke completed by the piston at the point of cut-off, and the total length of the stroke. For example, if the length of stroke is 48 inches, and the steam is shut off from the cylinder just as the piston has completed 15 inches of the stroke, the apparent cut-off is $\frac{15}{48} = \frac{5}{16}$.

The **real cut-off** is the ratio between the volume of steam in the cylinder at the point of cut-off and the volume at the end of the stroke, both volumes including the clearance of the end of the cylinder in question. If the volume of steam in the cylinder, including the clearance, at the point of cut-off is 4 cubic feet, and the volume, including the clearance, at the end of the stroke is 6 cubic feet, the real cut-off is $\frac{4}{6} = \frac{2}{3}$.

The **ratio of expansion,** also called the **real number of expansions,** is the ratio between the volume of steam, including the steam in the clearance space, at the end of the stroke, and the volume, including the clearance, at the point of cut-off. It is the reciprocal of the real cut-off. For example, if the volume at the end of the stroke is 8 cubic feet, and at the cut-off is 5 cubic feet, the ratio of expansion is $\frac{8}{5} = 1.6$; in other words, the steam would be said to have one and six-tenths expansions. The corresponding real cut-off would be $\frac{5}{8}$.

Let e = real number of expansions;
i = clearance, expressed as a per cent. of the stroke;
k = real cut-off;
k_1 = apparent cut-off;
r = apparent number of expansions $= \frac{1}{k_1}$.

Then,
$$e = \frac{1}{k} \text{ and } k = \frac{1}{e} \quad (1)$$
$$k = \frac{k_1 + i}{1 + i} \quad (2)$$

EXAMPLE.—The length of stroke is 36 inches; the steam is cut off when the piston has completed 16 inches of the stroke; the clearance is 4 per cent. Find the apparent cut-off, the real cut-off, and the real number of expansions.

SOLUTION.—Apparent cut-off $= \frac{16}{36} = \frac{4}{9} = .444$. Ans.

$$\text{Real cut-off} = k = \frac{k_1 + i}{1 + i} = \frac{.444 + .04}{1 + .04} = \frac{.484}{1.04} = .465. \quad \text{Ans.}$$

$$\text{Real number of expansions} = e = \frac{1}{k} = \frac{1}{.465} = 2.15. \quad \text{Ans.}$$

EXAMPLES FOR PRACTICE

1. Length of stroke, 18 inches; apparent cut-off, .4; clearance, 7.5 per cent. Find: (*a*) real cut-off, (*b*) real number of expansions.

Ans. { (*a*) .442
(*b*) 2.262

2. Length of stroke, 66 inches; clearance, 4 per cent.; steam cuts off at $14\frac{1}{2}$ inches. Find: (*a*) real and (*b*) apparent cut-off in per cent. of stroke; (*c*) real and (*d*) apparent number of expansions.

Ans. { (*a*) 24.97 per cent.
(*b*) 21.97 per cent.
(*c*) 4, nearly
(*d*) 4.552, nearly

CORLISS VALVE GEAR

DESCRIPTION

32. The **Corliss valve gear,** which is used in a large number of engines, differs from the plain slide valve in many particulars. In Fig. 21 is shown a side elevation of this valve gear, and in Fig. 22 a section through the cylinder and and valves. It has four separate and distinct valves. Two of these v, v', Fig. 22, connect directly with the steam chest d and steam pipe s, and are called *steam valves;* they are rigidly connected with the cranks n, Fig. 21, the right-hand crank being removed in order to show more clearly the disengaging hook i. The other two valves r, r', Fig. 22, connect directly with the exhaust chest l and the exhaust pipe o, and are called *exhaust valves;* they are rigidly connected with the cranks m, m, Fig. 21. All the valves are cylindrical in form, and extend across the cylinder above and below, respectively. The wristplate w is made to rock on a stud a, by the hook rod c, connecting it with an eccentric on the crank-shaft.

Two motion rods e, e connect the wristplate w with the bell-cranks h, h of the steam valves, and two motion rods f, f

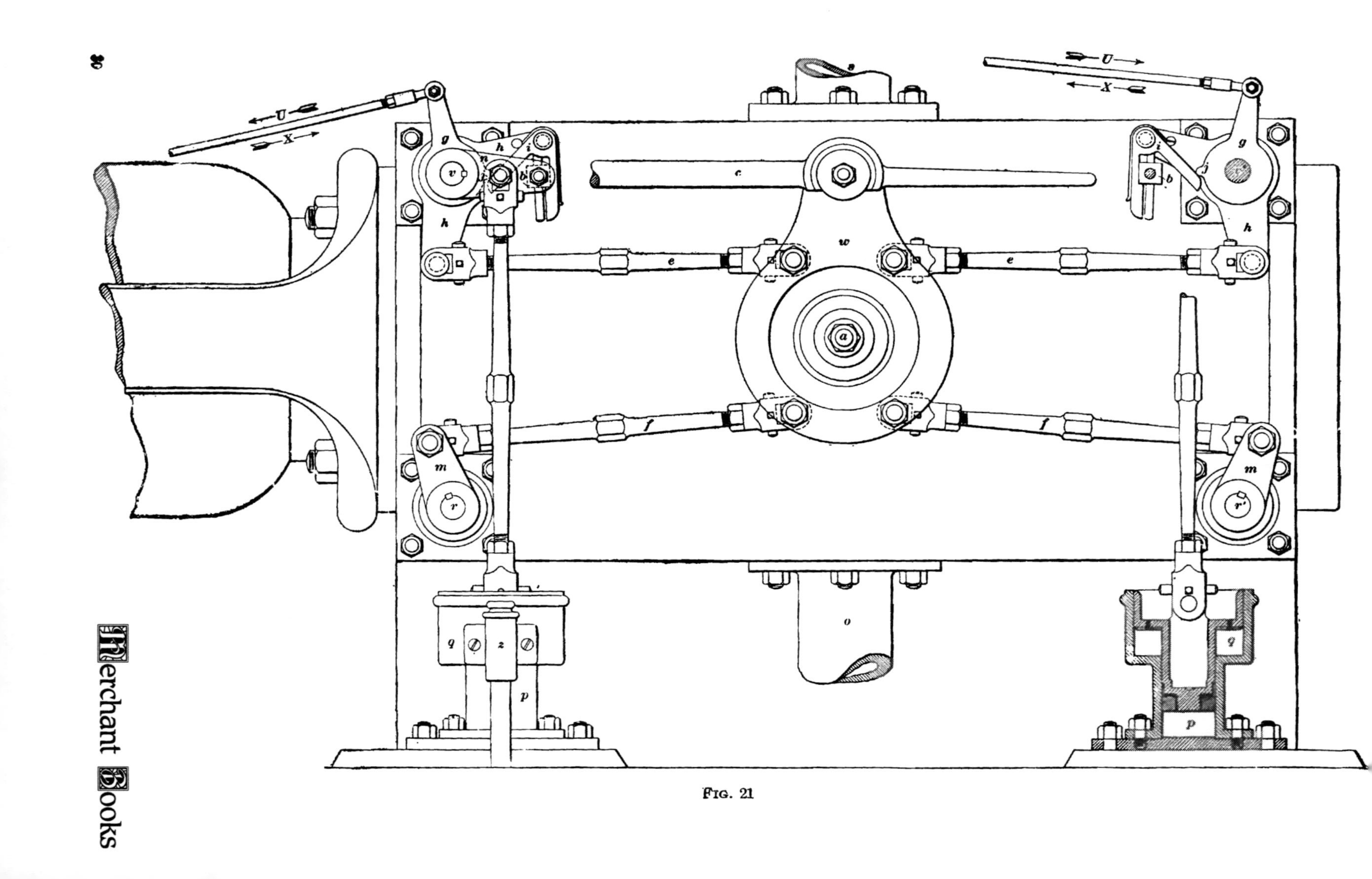

Fig. 21

connect the wristplate with the cranks m, m of the exhaust valves. The motion rods can be lengthened or shortened as the case may require, and the action of any one valve regulated independently of the other three. As the wristplate w rocks backwards and forwards, the exhaust valves r and r', which are rigidly connected with their cranks m, m, rock with it. The bell-cranks h, h, which are provided with the disengaging hooks i, i, are also given this rocking motion, and by hooking on to the blocks b, b, which are rigidly connected to the cranks n, open the steam valves v, v'.

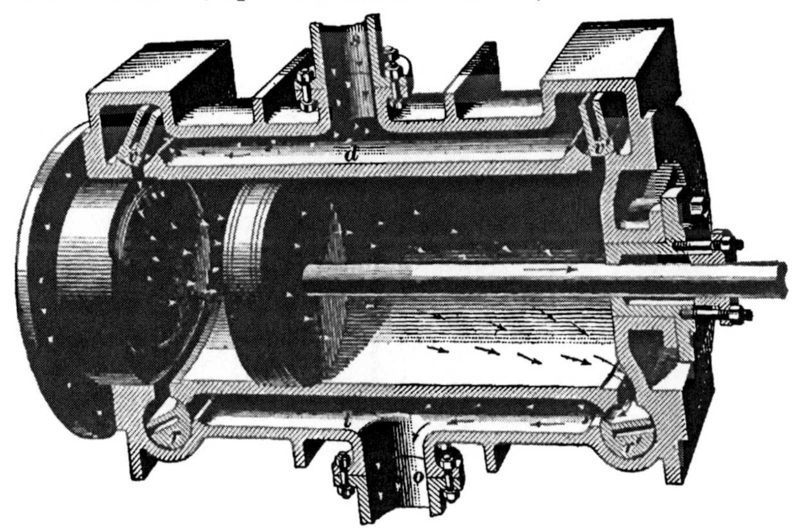

Fig. 22

The projections j, j on the two trip collars g, g unhook the disengaging hooks i, i, after they have rotated the valves v, v' through a certain angle, and the cranks n, n are pulled back to their first positions by the vacuum dashpots p, p, against the resistance of which the valve cranks n were raised. The governor changes the point of cut-off by moving the reach rods U, X, which are connected to the trip collars, thus enabling the projections j, j to be moved into various positions, causing the hooks i, i to disengage at any desired points. The movements of the valves open and close the steam and exhaust ports of the cylinder at the proper intervals. The pins of the motion rods are so located on the wristplate that the steam valves v, v' have their quickest movement while the exhaust valves r, r' have their slowest movement, and the exhaust valves have their quickest movement while the steam valves have their slowest movement. As a consequence of this arrangement, the steam and exhaust valves have entirely independent movements and the inlet ports may be suddenly opened full width by the quick movement of the steam valves, while the exhaust valves are practically motionless. The advantage of this valve gear is that it permits an earlier cut-off, a greater range of cut-off, a more perfect steam distribution, and a smaller clearance space than is attained with a plain slide valve.

Engines fitted with the Corliss valve gear do not usually run at much more than 100 revolutions per minute.

RELATIVE MOTIONS OF PISTON, CRANK, AND VALVES

33. Fig. 23 shows the piston nearing the end of its

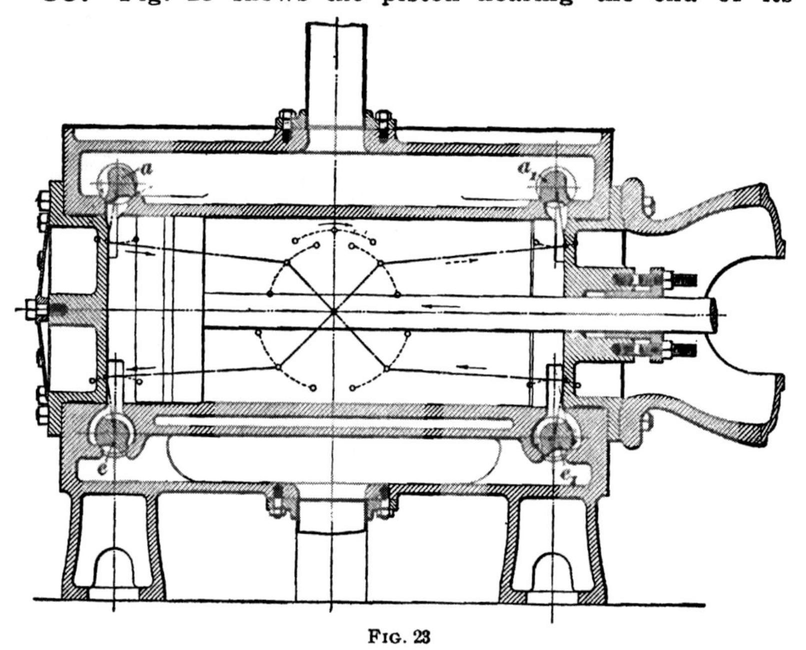

Fig. 23

return stroke and all the valves closed. The wristplate is in its middle position; hence, the exhaust valves are in

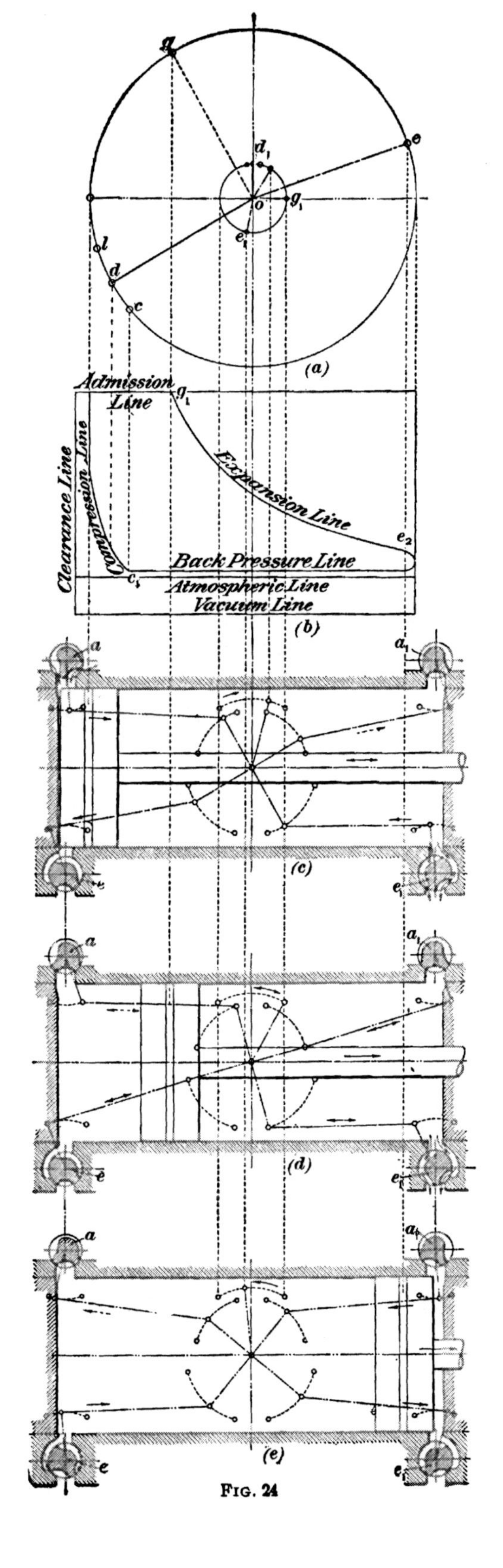

g
d₁
e
o
g₁
e₁
l
d
c
(a)
Admission Line
g₁
Clearance Line
Compression Line
Expansion Line
e₂
Back Pressure Line
c₁
Atmospheric Line
Vacuum Line
(b)
a
a₁
(c)
e₁
a
a₁
(d)
e
e₁
a
a₁
(e)
e
e₁

FIG. 24

precisely the same position, relative to their respective ports as would also be the two admission valves were it not that the one a_1 is in a released condition. Observing the arrows on the various motion rods, it will be seen that the exhaust valve e_1 will soon be opened to liberate the steam that is still exerting a pressure in the direction of motion of the piston, while the other valve e has just been closed; the admission valve a will also soon be opened to admit steam against the motion of the piston.

34. The diagrams in Fig. 24 (*a*), (*b*), (*c*), (*d*), and (*e*) show the most important simultaneous positions of the piston and the four valves, together with a skeleton outline of the principal members of the mechanism in their various corresponding positions, during a little more than a complete forward stroke of the engine.

In all the diagrams, the directions of motion of the piston, wristplate, and valve rods are indicated by arrows, a double-headed arrow being shown when a member is in the position in which its direction of motion is being reversed, and a dotted arrow being shown when an admission-valve-motion rod is moving without affecting its valve, the connecting parts being released.

In Fig. 24 (*a*), the respective positions of the crank and eccentric that correspond to the positions in Fig. 23 are d and d_1, and the position c of the crank indicates when the closing of the exhaust valve e actually took place and the resulting compression commenced, as indicated at c_1 in diagram (*b*), while the position l indicates where the steam valve a will open—in other words, the lead position of the crank. Diagram (*c*) shows the piston at the end of its return stroke, or, what is the same thing, at the beginning of its forward stroke. By this time, both the exhaust valve e_1 and the admission valve a have been opened considerably, without, however, reaching the limits of their opening positions, while the exhaust valve e has nearly reached the limit of its closing position, and this because the four points of attachment of the valve rods to the wristplate are, as will be seen, so located that the angular closing movements of the valves are very small compared with their angular opening movements.

35. In diagram (d), the wristplate is shown in the position in which its motion is just being reversed by the eccentric, the valve rods, in consequence, being also in the positions in which their motions are reversed, as indicated by the double-headed arrows. According to what has already been learned, this is the limiting position at which the admission valve a is released, which for that reason is supposed to have just occurred, as indicated by its closed position. As indicated at g_1, in diagram (b), this is then the moment at which in this particular case the expansion of the steam commences. The exhaust valve e has at the same time reached the limit of its closing position, while e_1 has reached the limit of its opening position. In this position of its motion rod, the admission valve a_1 is picked up, as indicated by the double arrow. In diagram (a), g and g_1 are, respectively, the positions of the crank and the eccentric that correspond to the valve positions of diagram (d).

36. In diagram (e), the piston is represented in the position near the end of its forward stroke, at which the exhaust valve e just begins to open, e_1 having been closed some time previously to produce compression, and the admission valve a_1 nearing its opening point.

Point e_2, in diagram (b), shows the release of the expanded steam due to the opening of e, and points e and e_1, in diagram (a), represent, respectively, the positions of crank and eccentric corresponding to diagram (e).

Merchant Books

STEAM-ENGINE DESIGN

(PART 1)

DATA AND CALCULATIONS

PRELIMINARY DATA

1. The designer of an engine has for his preliminary data: (1) The class of service for which the engine is intended, whether it is for a special case, as marine, mill, electrical, locomotive, hoisting, etc., or whether it is to be put on the market and sold wherever possible, to cover a large variety of classes; (2) the rated indicated horsepower; (3) the necessary economy; (4) the allowable fluctuation of speed; (5) occasionally, the boiler pressure; and (6) the type of engine desired—that is, whether simple or compound, horizontal or vertical, high- or low-speed, Corliss or slide-valve gear, etc.—which depends to a considerable extent on the nature of the preceding data.

For the economical range of load, the designer must determine: (1) The boiler pressure, if not already known; (2) the back pressure; (3) the point of cut-off at rated indicated horsepower; (4) the approximate piston speed; (5) the clearance, and (6) the amount of compression to be employed.

In order to design a simple engine after having obtained the foregoing data, it will be necessary to draw a theoretical indicator diagram and determine the mean effective pressure; then, the proportions of the cylinder can be calculated, and the design of the other parts will readily follow. For a compound or triple-expansion engine, a more complicated process is necessary.

2. The Boiler Pressures for Different Types of Engines.—The **boiler pressure** for an engine to be designed may be fixed beforehand, if it is known that the engine is to have steam furnished by an existing boiler or set of boilers carrying a definite pressure. In case the boiler pressure is not known, then experience has fairly definitely determined, for each type of engine, a range of pressure outside of which it is not desirable to go. This range is about as follows:

Type of Engine	Gauge Pressure, in Pounds
Simple	70 to 120
Compound	100 to 150
Triple-expansion	150 to 200 or higher
Quadruple-expansion	200 or higher
Locomotive	160 to 210

The ranges just given represent the best practical results for all conditions under which a steam engine is to operate, the chief consideration usually being the number of cylinders in series in which the expansion of the steam is to take place. As to choice of pressure within any of the ranges indicated, a low steam pressure is desirable, if simplicity and low first cost are the prime considerations; but if economy in weight and space are needed, a high steam pressure is necessary. The best all-around economy in fuel and other running expenses for any particular type of engine will usually be obtained with a steam pressure about the middle of the range for that type of engine.

The initial pressure in the engine cylinder will be less than the boiler pressure, on account of the loss caused by resistance to flow through the steam pipe and connections. Ordinarily, the loss may be taken at about 8 per cent. of the boiler pressure.

Even if the engine is rated on the basis of a lower pressure, the parts should be designed to carry safely a pressure of at least 100 pounds per square inch, gauge, with a back pressure as low as that ordinarily attained by a condenser, say 2 pounds per square inch, absolute; that is, with an unbalanced pressure of 112.7, say 113, pounds per square inch, for the engine may at some time be run with a condenser, and, when starting, stopping, or running at low speed, the unbalanced pressure may become equal to the full boiler pressure plus the condenser pressure.

3. Piston Speed.—The best practical results may be attained by using the following **piston speeds:**

Type of Engine	Piston Speed, in Feet per Minute
Stationary, small	300 to 600
Stationary, medium size	600
Stationary, large	750
Marine	850 to 900
Locomotive	600 to 1,200

4. Economical Ratio of Expansion.—In a simple engine, the greatest economy of steam occurs with a **ratio of expansion** varying from about 3 to 5, and within this range of the ratio of expansion, the *steam economy* does not vary much. It is customary to rate an engine at its most economical load, which, in the case of a simple engine, will occur with a ratio of expansion between 3 and 5, except with the plain slide-valve engine having no cut-off mechanism. As it is more common to run a steam engine underloaded than overloaded, most of the medium- and high-speed automatic cut-off engines constructed at the present time by American builders are rated on a ratio of expansion of about 3, with the expectation that the engine will usually run inside of its economical range. It should be remembered that the ratio of expansion mentioned here is the true ratio of expansion; and this, at a value of 3, with the large clearance used in high-speed automatic cut-off engines, brings the cut-off at about .25 stroke at rated load.

As the latest cut-off of single-wristplate Corliss engines is between .4 and .5 stroke, the makers of such engines usually rate them on cut-off at .2 stroke, in order to give them some overload capacity. This makes the true ratio of expansion of such engines at rated load between 4 and 5, with the usual clearance.

5. For a compound engine, the total combined theoretical indicator diagram may be drawn, and this may be divided between the cylinders. The division of the diagram shows the proper ratios of expansion for each cylinder. If the maximum of steam economy is desired, the diagram should be divided so as to give about an equal range of temperature to each cylinder. If the greatest uniformity of rotative speed is desired, the diagram should be divided so as to give about the same amount of work to each cylinder. If both are desirable, an intermediate division of the diagram becomes necessary. A simple extension of the method gives the ratios of expansion for triple- and quadruple-expansion engines.

6. Clearance.—The term *clearance* in connection with the steam engine is used in two senses. The clearance may be the distance between the piston and the cylinder head when the piston is at the end of its stroke, or it may represent the volume between the piston and the valve when the engine is on dead center. To avoid confusion, the former is called **piston clearance,** and the latter is simply termed **clearance.** Piston clearance is always a measurement expressed in parts of an inch. Clearance, however, is a volume, and it is usually taken as a percentage of the volume swept through by the piston per stroke. Thus, a clearance of 7.5 per cent. signifies that the volume just defined is 7.5 per cent. of the volume swept through by the piston during one stroke. The usual values of the clearance in various types of engines as found in practice are:

Type of Engine	Clearance Per Cent.
Four-valve, experimental	1
Corliss and other drop cut-off	1.5 to 3.5
Medium-speed	3 to 8
High-speed, with long slide valves	4 to 12
High-speed, with short, plain slide valves	7 to 15

Within the ranges just stated, the large engines of each type will have the smaller clearances, and the small engines will have the larger clearances. High rotative speed tends to large clearance, and low rotative speed to small clearance. In slide-valve engines, piston valves give larger clearances than flat valves.

It is usually desirable to reduce the clearance to a minimum. However, with the early exhaust closure of single-valve automatic engines, large clearance reduces the pressure in the cylinder at the end of compression below what might otherwise be objectionable; and, with high-speed engines, the large clearance also reduces the danger due to water in the cylinder.

7. When used as a linear distance, clearance is usually spoken of as *piston clearance* or, sometimes, *mechanical clearance;* it is the shortest distance in the direction of the stroke between the piston when at the end of the stroke and the nearest cylinder head, and should be made as small as possible. On small stationary engines, this distance may be $\frac{1}{4}$ inch, and it rarely exceeds $\frac{1}{2}$ inch on the largest marine engines. In some cases in actual practice, with low-pressure cylinders 7 feet in diameter and conical pistons, this clearance is only $\frac{3}{8}$ inch.

For the purpose of design the piston clearance may generally be found by the formula:

$$c = a + b\,x,$$

in which c = piston clearance, in inches;

x = number of bearings, or joints, between piston and crank-shaft, where there can be play;

a = $\frac{1}{8}$ inch for engines of 35 horsepower or less; $\frac{3}{16}$ inch for engines between 35 and 100 horsepower; and $\frac{1}{4}$ inch for engines above 100 horsepower;

b = $\frac{1}{32}$ inch for engines of 35 horsepower, or less; $\frac{3}{64}$ inch for engines between 35 and 175 horsepower; and $\frac{1}{16}$ inch for engines above 175 horsepower.

The horsepowers just stated are taken at rated load; that is, the load for which the engine was designed. For very fast-running engines the allowance for each bearing may be doubled.

ENGINE CALCULATIONS

BACK PRESSURE AND POINT OF EXHAUST CLOSURE

8. Back Pressure and Compression.—In a well-designed non-condensing engine, the **back pressure** should not exceed 16 or 17 pounds per square inch, absolute. For a condensing engine, the back pressure may be from 2 to 4 pounds per square inch, absolute.

The method of finding the proper amount of compression depends on the class of valve gear used, since the point of exhaust closure may remain the same at all loads, as in the Corliss or other independent cut-off valve gear; or, the exhaust closure may change with the load, as in the single swinging-eccentric gear controlled by a shaft governor.

The case in which the point of exhaust closure does not change will first be considered. Taking the ratio of expansion at rated load appropriate to the case, as explained in Art. 4, and the expansion curve as pv = a constant, then the pressure at the end of the stroke will be the initial absolute steam pressure in the cylinder divided by the ratio of expansion.

The next step is to find the weight of the piston, piston rod, and crosshead. This weight may be estimated with all necessary accuracy by examining the records of weights of these parts in engines of similar type and, as near as possible, the same power. A designer usually has access to such data in the office records of his employer.

Let w = weight of reciprocating parts, that is, sum of weights of piston, piston rod, crosshead, and half the connecting-rod, in pounds;

N = number of revolutions per minute;

r = length of the crank, in feet;

F = total inertia pressure at end of stroke, in pounds.

The reciprocating parts, on account of their inertia at the end of the stroke, will exert a pressure in the direction of motion. This inertia pressure may be expressed in pounds with sufficient accuracy by the following formula, which is based on the assumption that the connecting-rod is of infinite length and is explained in *Mechanics of the Steam Engine*.

$$F = .00034\,w\,N^2\,r \qquad (1)$$

Let A = piston area, in square inches, taking an approximate value from the records, as just explained;

F_1 = inertia pressure at end of stroke, in pounds per square inch of piston area.

Then, formula **1** becomes

$$F_1 = \frac{.00034\,w\,N^2\,r}{A} \qquad (2)$$

Let p_2 = absolute forward steam pressure at the end of the stroke, in pounds per square inch; that is, the initial pressure divided by the ratio of expansion;

p' = total forward absolute pressure on the piston at the end of the stroke, in pounds per square inch.

Then, $$p' = p_2 + F_1 \qquad (3)$$

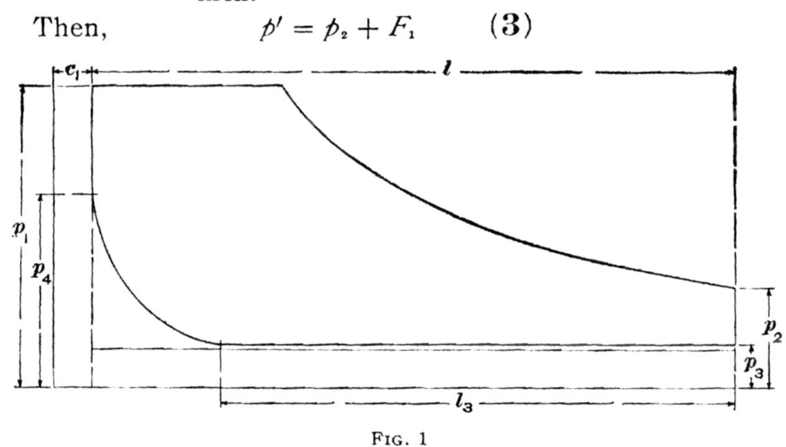

Fig. 1

9. Point of Exhaust Closure.—In Fig. 1 is shown a diagram that indicates the pressures and percentages of stroke. The initial pressure is p_1, the pressure at the end of expansion p_2, the back pressure p_3, and the compression pressure p_4, all absolute pressures in pounds per square inch. The length of the stroke is l, the part of the stroke from the beginning of exhaust to the point where the exhaust valve closes is l_3, and the clearance is c_1. As l_3 and c_1 are usually expressed as percentages of the stroke, then l, or 100 per cent. of the stroke, is expressed as 1.

To cushion the reciprocating parts properly, the pressure at the end of the stroke due to compression should equal p'. Remembering that the volumes in the cylinder are proportional to the percentages of the lengths of stroke, the pressure p' times the clearance c_1 equals the absolute back pressure p_3 at exhaust closure multiplied by the clearance plus the portion of the stroke yet to be completed, which is $(1 + c_1 - l_3)$, or, expressed as a formula, $p' c_1 = p_3(1 + c_1 - l_3)$, which reduces to

$$l_3 = 1 - \frac{(p' - p_3) c_1}{p_3} \qquad (1)$$

Lacking the practical data that would have to be known to use the formula just given, a trial value of the pressure in the cylinder at the end of compression may be assigned by the use of the following empirical formula:

$$p_4 = (.0025 N + .2) p_1 \qquad (2)$$

in which p_1 = initial steam pressure in the cylinder, in pounds per square inch, absolute;

p_4 = absolute steam pressure in the cylinder at end of compression;

N = number of revolutions per minute.

Then, as p_4 should equal p', this value may be used in formula **1**, giving

$$l_3 = 1 - \frac{(p_4 - p_3) c_1}{p_3} \qquad (3)$$

10. After the design of an engine has proceeded far enough, the calculation based on the weight of reciprocating parts can be made from the results obtained. If $p' = p_4$, that is, if the total forward pressure on the piston at the end of the stroke equals, approximately, the back pressure due to compression at the end of the stroke, the result will be sufficiently accurate for present use; but if these two quantities are very much different, such modifications should be made as may be necessary to bring them approximately equal, provided the engine is one of high or medium speed.

With low-speed engines, such as the Corliss, it is usually unnecessary to pay much attention to the equality of the steam pressure at the end of compression, with the total forward pressure on the piston. In these, if the exhaust closure takes place a little before the admission of live steam at the same end, say at 95 per cent. of the stroke, the results will be satisfactory for all practical purposes.

11. Exhaust Closure in Engine With Single Swinging Eccentric.—With a single swinging eccentric a different method must be followed. It may be assumed that the power will not be sufficient to run the engine against its own friction with a cut-off earlier than $\frac{1}{20}$ stroke, and that the engine will therefore never run with a cut-off earlier than this. Then, at this earliest cut-off, the back pressure at the end of the stroke due to compression should not exceed the initial steam pressure in the cylinder; otherwise, the valve may be forced from its seat. To avoid this, the following formula, in which the values of the different quantities are the same as in Art. 9, should be used:

$$l_3 = 1 - \frac{(p_1 - p_3)c_1}{p_3} \qquad (1)$$

This formula gives the percentage of the stroke at which exhaust closure should occur at the earliest cut-off. Then, by means of the valve diagram, the resulting point of exhaust closure may be found for the rated load and ratio of expansion. This will be illustrated by an example in the succeeding article.

When only a rough approximation is required, the following formula is sometimes used:

$$p_4 = \frac{p_1 + 16}{2} \qquad (2)$$

FUNDAMENTAL ENGINE CALCULATIONS

12. Calculations for Simple Non-Condensing Engine.—The general method of procedure in a design will be illustrated in the following example:

Let it be required to determine the diameter of cylinder, length of stroke, and number of revolutions of a simple, moderate-speed, non-condensing engine, having independent cut-off, slide-valve gear, and fixed compression, to develop 100 indicated horsepower at the rated load.

As there is no particular reason in this case for going to an extreme in steam pressure, a pressure in about the middle of the range for a simple engine will be taken; namely, 100 pounds gauge pressure or 114.7 pounds, absolute. Then, assuming a ratio of expansion of 3, the terminal pressure in the cylinder will be $114.7 \div 3 = 38.2$ pounds, absolute.

Suppose that the designer has access to the records and drawings of two engines, similar to the one in hand, on which appear the following data:

Rated Horsepower	Weight of Reciprocating Parts	Area of Piston	Length of Stroke
	Pounds	Square Inches	Inches
60	450	78.5	18
130	542.8	159.48	22

It will be sufficiently accurate for the purpose to assume that the weight of the reciprocating parts of the proposed engine may be found by interpolating between the weights of the two engines just given. This may be done by comparing the difference in horsepower with the difference in weight. The difference in the horsepowers of the known engines is $130 - 60 = 70$, and that in the weights is $542.8 - 450 = 92.8$. Hence, for a difference of 70 horsepower, there is a difference of 92.8 pounds in the weights of the reciprocating parts. The difference between 100 horsepower and 130 horsepower is 30 horsepower. The difference in weight corresponding to this difference in horsepower is $\frac{30}{70} \times 92.8$, or 39.8 pounds. This gives the weight of the reciprocating parts for an engine of 100 horsepower as 542.8 – 39.8 = 503 pounds.

Similarly, the area of the piston will be the larger area less $\frac{30}{70}$ of the difference between the given areas, or 159.48 – $\frac{30}{70}$ (159.48 – 78.5) = 124.77, say 124.8, square inches.

13. In the course of a mechanical design, it often becomes necessary to estimate or assume approximate values before final values can be calculated. This can be done most satisfactorily by the foregoing process of interpolation from existing designs. Such interpolation does not usually enable the designer to determine values with enough accuracy to use them as working dimensions, but it is of great service in enabling him to reach a working approximation to values that enter into the calculations. In the absence of information from previous designs, necessary preliminary estimates must be made by unaided judgment; then, designing becomes largely a trial process, in which the calculations must be made and remade until the final result is satisfactory.

By interpolating again, the approximate length of stroke of the proposed engine is found to be $22 - \frac{30}{70}(22 - 18) = 20\frac{2}{7}$ inches. As this is an inconvenient dimension, 20 inches may be taken. Then, r, the length of the crank in feet, is 10 ÷ 12. For a piston speed of 600 feet per minute, from Art. **3,**

$$N = \frac{600 \times 12}{20 \times 2} = 180 \text{ revolutions per minute}$$

From the values thus found, the pressure of the reciprocating parts at the end of the stroke may be found by means of formula **1,** Art. **8;** thus, $F = .00034 \times 503 \times (180)^2 \times \frac{10}{12} = 4{,}617.54$ pounds, and the corresponding pressure per square inch of piston is 4,617.54 ÷ 124.8 = 37 pounds. The total forward pressure per square inch of piston, at the end of the stroke, is then 38.2 + 37 = 75.2 pounds.

14. It is now necessary to assign a value to the clearance. Referring to Art. **6,** the small size of the engine as well as the form of valve gear would indicate that the clearance must be large, while the moderate rotative speed would tend to give small clearance. Hence, a value well up in the range for medium-speed engines, but not at the very top, may be assigned to the clearance. Let a clearance of 6 per cent. be assumed, and, from Art. **8**, a back pressure of 17 pounds, absolute. Then, the necessary point of exhaust closure is at the fraction of the stroke given by formula **1**, Art. **9**; thus,

$$l_s = 1 - \frac{.06\,(75.2 - 17)}{17} = .795, \text{ or } .8, \text{ nearly}$$

15. There is now sufficient data to draw the theoretical diagram, as shown in Fig. 2, the scale to which the diagram

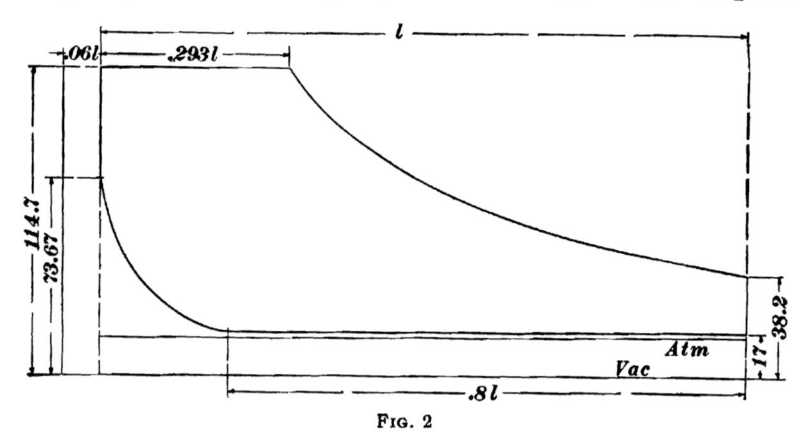

FIG. 2

should be drawn being entirely a matter of convenience. Let the pressure scale in this case be 50 pounds to the inch and the volume scale such that the diagram is 5 inches long; then, the clearance is $5 \times .06 = .3$ inch.

The diagram may now be drawn, assuming that the expansion and compression curves follow the law pv = a constant. By either measurement or calculation, the *mean effective pressure*, or the M. E. P., of the diagram, Fig. 2, is found to be 54 pounds per square inch. On account of the cylinder condensation and other losses, the M. E. P. given by the theoretical card is never attained by the actual engine. To find the probable M. E. P. of the actual engine, the M. E. P. of the theoretical card must be multiplied by a factor, the magnitude of which depends on the type of engine.

For jacketed engines, with Corliss valve gear or other quick-acting drop cut-off gears, or multiple-ported valves with good-sized ports for swinging eccentric engines, a factor of .94 is used.

For jacketed engines, with slide valves, either single-ported with large travel, large ports, and late cut-off, or double-ported valves in which the port opening in the early part of the stroke is not particularly good, a factor of .9 to .92 is used.

For unjacketed engines, with plain **D**-slide valve for cut-off about one-half stroke or earlier, or where the port opening is rather small in order to secure small travel with an unbalanced valve, a factor of .8 to .85 is used.

The foregoing values of the diagram factor are taken from Seaton, and represent conservative practice. Many American designers use factors somewhat higher than these, and employ the same factor for both jacketed and unjacketed engines of any particular type. This is done because it is supposed that the jacket affects the quantity of steam taken into the cylinder, rather than the area of the indicator diagram.

In the case of the simple engine of 100 horsepower under consideration, the factor will be about .94; then, the probable M. E. P. is $54 \times .94 = 50.76$ pounds per square inch.

16. The formula for the indicated horsepower of an engine is

$$\text{I. H. P.} = \frac{PLAN}{33{,}000} \qquad (1)$$

in which I. H. P. = indicated horsepower;
P = mean effective pressure, in pounds per square inch;
L = length of stroke, in feet;
A = area of piston, in square inches;
N = number of revolutions per minute.

The area of the piston or cylinder may be found by transforming formula **1**; thus,

$$A = \frac{33{,}000 \times \text{I. H. P.}}{PLN} \qquad (2)$$

As the product of the length of the stroke L and the number of revolutions N equals the piston speed in feet per minute, the piston speed may be substituted for $L N$ in the formula. Then, in the case of an engine of 100 horsepower and a piston speed of 600 feet per minute, formula **2** gives

$$A = \frac{33{,}000 \times 100}{50.76 \times 600} = 108.4 \text{ square inches}$$

The area A just found is sufficiently accurate for practical purposes as a preliminary estimate; however, it must be taken as the average area of the piston. As the effective area of one side is diminished by the area of cross-section of the piston rod, the area of the other, or free, side of the piston must be increased accordingly.

It will be accurate enough for the purpose to make this correction in the following manner:

Let D = diameter of the cylinder, in inches;
A = area of the cylinder, in inches.

Then, $A = .7854\,D^2$, or $D = \sqrt{A \div .7854}$

Hence, in this case,

$$D = \sqrt{108.4 \div .7854} = 11.75 \text{ inches,}$$

which is near enough for the present purpose.

Let d = diameter of the piston rod, in inches;
p = maximum unbalanced pressure on the piston, in pounds per square inch.

Then,

$$p = 114.7 - 2 = 112.7, \text{ say } 113, \text{ pounds}$$

The diameter of the piston rod may be found approximately by the formula

$$d = .02\,D\sqrt{p}$$

from which

$$d = .02 \times 11.75 \times \sqrt{113} = 2.5 \text{ inches}$$

The area of a circle of this diameter is 4 9 square inches.

Let A be the area of the free side of the piston; then,

$$A = 108.4 + \frac{4.9}{2} = 110.85 \text{ square inches,}$$

from which, the required diameter of the piston and cylinder is 11.89 inches.

If the design is one of a line of engines to be built in large numbers and in standard sizes, the diameter of the cylinder would probably be taken as 12 inches. The effect of this change to even dimensions will be that, at the rated load, the ratio of expansion will differ somewhat from 3; but there is no objection to this slight departure from the ratio of expansion assumed in the design. The other values found are also within the range of good practice and may therefore be retained.

17. Calculations for High-Speed Automatic Cut-Off Engine.—As a further illustration, let it be required to find the diameter of cylinder, length of stroke, and rotative speed of a simple, high-speed, non-condensing engine of 100 I. H. P. at rated load, with swinging eccentric, automatic cut-off, and double-ported flat slide valve of moderate length.

By referring to Art. **6,** it will be reasonable to expect a clearance of about 10 per cent. in this case, and, as before, an initial pressure of 100 pounds, gauge, in the cylinder may be assumed. From Art. **11,** it may be assumed that the cut-off will not occur earlier than one-tenth stroke. As the boiler pressure is 100 pounds, gauge, the initial pressure p_1 will be 114.7, and, according to Art. **8,** the back pressure may be taken as 17 pounds, absolute.

Applying formula **1,** Art. **11,** with cut-off at one-tenth stroke, the exhaust must close not earlier than

$$l_3 = 1 - \frac{.1\,(114.7 - 17)}{17} = .425 \text{ of the stroke}$$

With a ratio of expansion of 3, from Art. **4,** and a clearance of 10 per cent., the cut-off will come at $\frac{1.1}{3} - .1 = .27$ of the stroke, nearly.

Then, by the provisional valve diagram, as shown in Fig. 3, the exhaust closure is found to be at .562 stroke when the cut-off is at .27 stroke. The use of the Bilgram valve diagram, as explained in *Valve Gears*, Part 1, is recommended for this purpose.

The line $A C$, Fig. 3, is first drawn to represent the stroke of the engine, and the semicircle $A B C$, with O as its center, is drawn to represent the crank-circle. Then, .1 l is laid off to represent the earliest cut-off position of the piston. The perpendicular a and the radial line $O D$, representing the crank position at earliest cut-off, are then drawn. The angle $D O C$ is bisected by the line $O E$, and any point, as O_1, on this bisector is taken as a center for a circle b tangent to $D O$ and $O C$. Next, the earliest point of exhaust closure .425 l is laid off, the perpendicular c erected, and the radial line

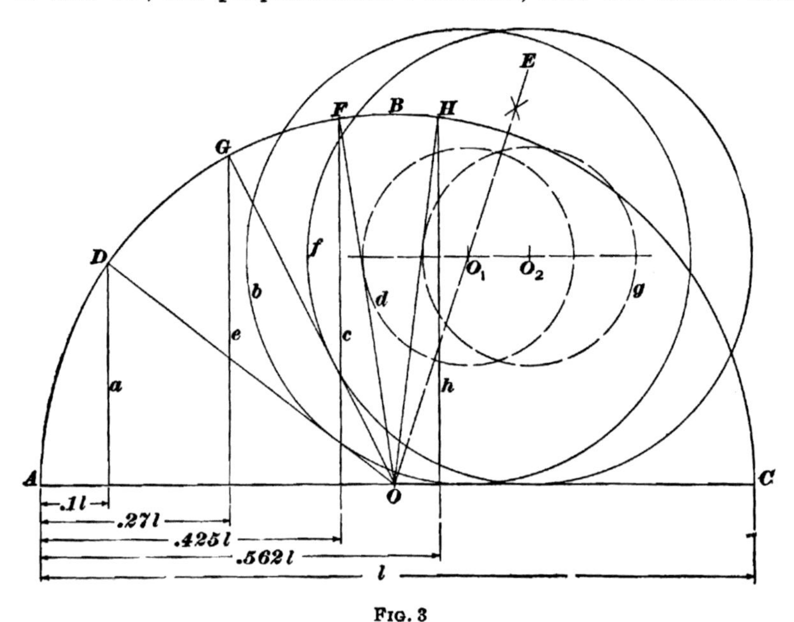

FIG. 3

$O F$ drawn to represent the crank position at exhaust closure. Then, with O_1 as a center, draw the circle d tangent to $O F$. Lay off .27 l on $A C$ to locate the point of cut-off for a ratio of expansion of three, erect the perpendicular e, and draw the crank-line $O G$. Then, on the horizontal line through O_1, locate O_2, so that a circle of the same radius as b will be tangent to $O G$ and $O C$, as circle f; draw circle g with O_2 as a center and of same radius as d, and draw $O H$ tangent to g. Then, drop the perpendicular h, and its distance from A will give .562 l, the point of exhaust closure. No values, except the point of exhaust closure at rated load, should be taken from the provisional valve diagram.

18. Now, from the theoretical indicator diagram, as in the preceding case, the M. E. P. is found to be 52.47 pounds, and the actual M. E. P. will be about $52.47 \times .94 = 49.3$ pounds. Then, by using formula **2**, Art. **16**, and remembering that $L\,N$ = piston speed, the area of the cylinder is

$$A = \frac{33{,}000 \times \text{I. H. P.}}{P\,L\,N} = \frac{33{,}000 \times 100}{49.3 \times 600}$$

$$= 111.6 \text{ square inches, nearly}$$

Correcting for the area of the piston rod, exactly as in the previous case, the area of the free side of the piston is found to be 114 square inches, from which the diameter of the piston and of the cylinder is found to be 12.05 inches. As in the previous case, the diameter would usually be taken as 12 inches.

As the engine is to run at high speed, about 250 may be taken as the number of revolutions per minute. Using this rotative speed and a piston speed of 600 feet per minute, the length of stroke becomes $\frac{600 \times 12}{250 \times 2} = 14.4$ inches. The stroke would then usually be made an even 14 inches, making the rotative speed $\frac{600 \times 12}{14 \times 2} = 257$ revolutions per minute.

19. Calculations for Hoisting and Locomotive Engines.—In the case of hoisting engines, locomotives, and other engines that must start a heavy load from rest, the required diameter of cylinder is determined by the *torque*, or *twisting moment*, on the shaft necessary to start the load. If the cylinder is made large enough to start the load, it will always run the ordinary load at an early cut-off and reduced power. In engines of this class, two cylinders are used.

In order to illustrate, take the case of a hoisting engine, the maximum torque on the shaft of which, in starting the load, is 149,000 foot-pounds. To secure ease in handling the load, stopping, starting, etc., such an engine should be of long stroke—two or three times the diameter of the cylinder—and of slow, rotative speed. Hence, the revolutions per minute may be taken as 65.

20. To insure perfect handling, the engine must have a minimum torque somewhat in excess of 149,000 foot-pounds. Hence, allow 20 per cent. for overcoming friction and accelerating the moving parts. Engines of this type have two cylinders and the cranks are connected at right angles. Thus, it may be assumed with sufficient accuracy, that the minimum torque of the engine will occur when one crank is on the dead center. By neglecting the angularity of the connecting-rod and assuming that the full boiler pressure acts on the piston, the minimum torque of the engine is then the product of the area of one piston, the steam pressure, and the length of the crank. The back pressure may be taken as that of the atmosphere, so that, by taking a boiler pressure of 100 pounds, gauge, the forward pressure in the cylinder will be 100 pounds per square inch of piston area.

Let D be the diameter of the cylinder in inches. Taking the length of the stroke as twice the diameter of the cylinder, the length of the crank in feet is $D \div 12$. Including friction, the torque to be overcome by the engine is $100 \times .7854\, D^2 \times D \div 12$. Equating these values with the starting load of 149,000 with 20 per cent. added gives $78.54\, D^3 \div 12 = 149{,}000 \times 1.2$, or

$$D = \sqrt[3]{149{,}000 \times 1.2 \times 12 \div 78.54} = 30.12, \text{ say } 30, \text{ inches}$$

Hence, the stroke is $30 \times 2 = 60$ inches.

21. Compound and Triple-Expansion Engines. The diameter of the low-pressure cylinder of a compound or triple-expansion engine may be found as in the foregoing cases by assuming that all the work is done in the low-pressure cylinder. In this case, the factor by which to multiply the theoretical M. E. P. to obtain the probable M. E. P. is from .7 to .8 for a compound and .6 to .7 for a triple-expansion engine. As the ratio of the volume of the high-, intermediate-, and low-pressure cylinders is determined, the diameter of the high- and intermediate-pressure cylinders may be found from that of the low-pressure cylinder.

ENGINE DETAILS

CYLINDERS AND STEAM CHESTS

22. Cylinder Proportions.—The proportions here given are largely empirical, being based on an examination of a large number of engines of the classes treated. Fig. 4 illustrates an example of a cylinder designed for a simple slide-valve engine. The crank-end head A is cast solid with the cylinder, while the method of fastening it to the frame B is clearly shown. The thickness of the cylinder walls may be found from either of the following formulas, according to the kind of work the engine is to do:

$$i = .0002 p D + .4 \text{ inch} \quad (1)$$

or,

$$i = .0003 p D + .375 \text{ inch} \quad (2)$$

in which i = thickness of walls of cylinder, in inches;

p = maximum steam pressure, in pounds per square inch, gauge.

D = diameter of cylinder bore, in inches; if p is less than 100 in the actual case, use 100 in the design.

Formula **1** applies only to engines in which lightness is a prime consideration, and in which first-class material and workmanship are assured by careful inspection and tests throughout the whole process of manufacture. Formula **2** applies to cylinders above 10 inches in diameter and gives a value of i much greater than is given by formula **1.** Formula **2** is of general application to ordinary cylinders of slide-valve engines. Cylinders in which the length is great in proportion to the diameter will be treated later.

23. In Fig. 4, the principal dimensions of the cylinder are indicated by letters whose value may be determined by the following formulas, in which D is the diameter of the cylinder in inches:

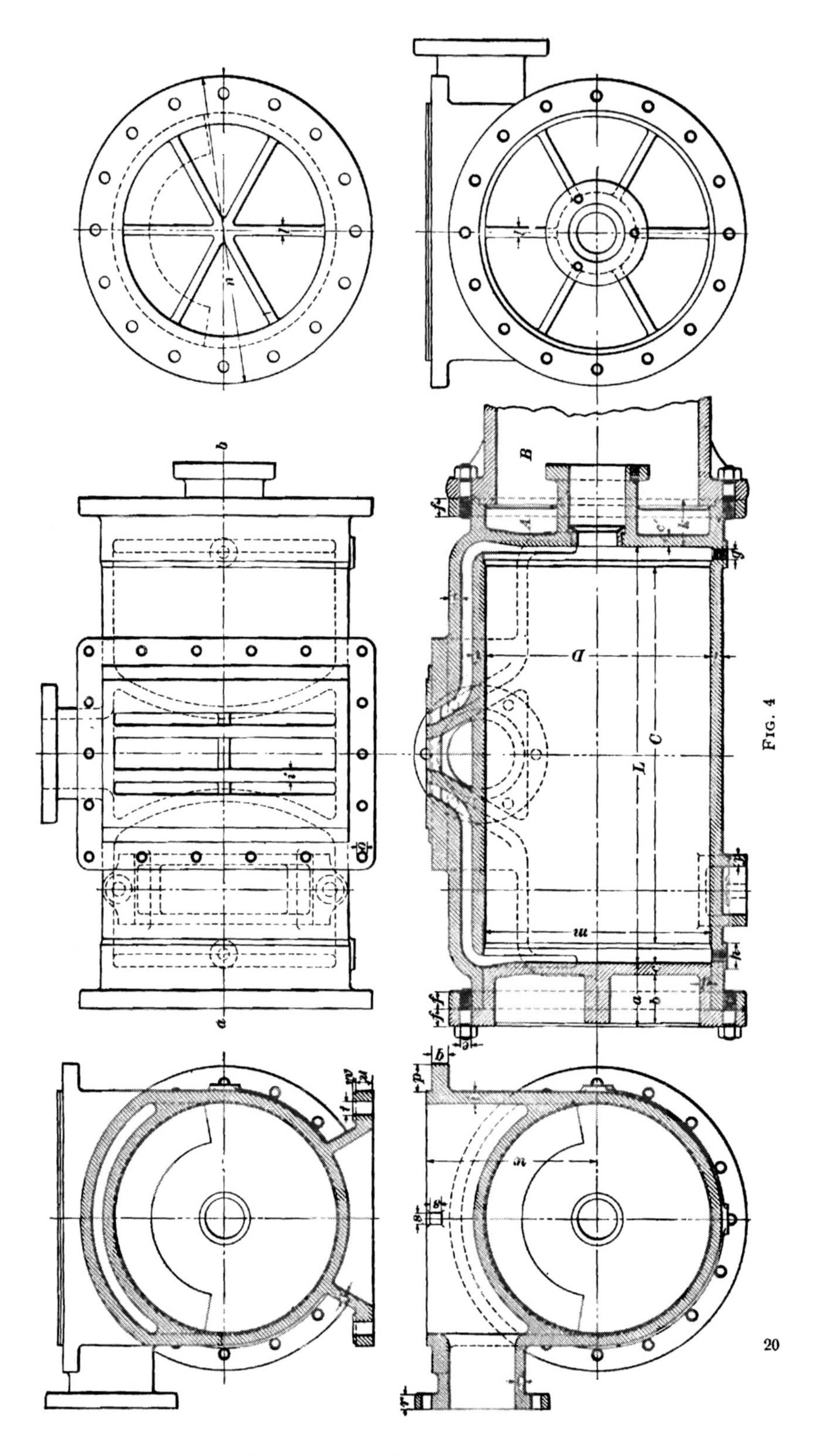

Fig. 4

L = length of stroke + thickness of piston + twice the piston clearance.

C = length between counterbores = length of stroke + distance from outer edge to outer edge of piston rings − $\frac{1}{3}$ width of piston ring.

$a = 5.5\,i$.

$b = 4.2\,i$.

$c = i + .25$ inch.

$c' = 1.1\,(i + .25$ inch).

$d = i$.

e = nominal diameter of cylinder-head bolt or stud. (The method of finding this value is given in the following article.)

$f = 1.5\,i$.

$g = .04\,D + .125$ inch. (Take the nearest nominal size of pipe tap.)

h = twice the outside diameter of drain pipe.

i is found by formula **1** or **2**, Art. **22**.

$j = .85\,i$.

$k = 4\,i$.

$l = .75\,i$.

$m = 1.01\,D + .125$ inch.

$n = m + 6\,e$.

o = nominal diameter of steam-chest cover bolts. (The method of finding this value is given in the following articles.)

$p = 2.75\,o$.

$q = 1.5\,r$.

$r = 1.25\,i$.

$s = i$. (This value is required only when the length of the ports is greater than 12 inches.)

$t = 1.25\,i$; when D is greater than 24 inches, use four bolts in the standard, and make $t = 1.1\,i$.

$u = 1.5\,i$.

$v = .25$ inch.

24. Cylinder-Head and Valve-Chest Cover Bolts. The strength of the cylinder-head and the valve-chest cover bolts or stud bolts holding a cylinder or valve-chest cover, over and above allowances due to screwing up, must be sufficient to hold the cover against ordinary boiler pressure; and, at the same time, the number of bolts must be great enough and close enough together to prevent springing of the flange between bolts, and consequent leakage. A large number of small bolts would tend to secure the latter result; but it must be remembered that the intensity of stress due to screwing up is much more severe in a small bolt than in a large one. Consequently, the bolts used are seldom less than $\frac{3}{4}$ inch in diameter, nor are they often larger than $1\frac{1}{4}$ or $1\frac{3}{8}$ inches in diameter in an engine of ordinary size. The usual range is from $\frac{7}{8}$ to $1\frac{1}{4}$ inches in diameter. Ordinarily, a sufficient number of bolts will be used, so that they will not be farther apart than about 5 inches from center to center around the bolt circle nor more than five times the thickness of the flange in which they are placed, as a steam-tight joint will not be obtained with spacing that is much wider.

The number of bolts is generally determined by convenience in spacing, an even number being used.

Let d = outside nominal diameter of bolt, in inches;
D_1 = diameter, in inches, of cylinder in counterbore;
l = load, in pounds, on each bolt, due to steam pressure;
L = load, in pounds, on each bolt, due to screwing up;
n = number of cylinder-head bolts;
p = maximum steam pressure, in pounds per square inch, gauge;
S_1 = safe tensile stress of material, in pounds per square inch.

The load on each bolt due to screwing up may be taken as

$$L = 16{,}000\,d \qquad (1)$$

The load on each bolt due to the steam pressure is expressed by the formula

$$l = \frac{.7854\,D_1^2\,p}{n} \qquad (2)$$

It will be on the safe side to take the total load on one bolt as the sum of the loads due to screwing up and the steam pressure, or $L + l = 16{,}000\,d + \frac{.7854\,D_1^2 p}{n}$. The safe tensile stress in one bolt is $.7854\,d^2 S_1$, approximately; hence, by placing this value equal to the total load on one bolt, and solving for d, the following formula is obtained:

$$d = \frac{10{,}186}{S_1} + \sqrt{\frac{D_1^2 p}{S_1 n} + \left(\frac{10{,}186}{S_1}\right)^2} \qquad (3)$$

In good practice a factor of safety of about 3 is commonly used. With small bolts, it may be necessary to allow a factor as low as 2.9, while with large bolts, a factor of about 3.3 may be obtained. This makes the safe tensile stress in the bolts from about 15,000 to 17,000 pounds per square inch for wrought iron, and from 18,000 to 21,000 for steel.

The nominal outside diameter of a bolt or stud of either wrought iron or steel should not be less than $\frac{3}{4}$ inch. Having found the required diameter by the foregoing formulas, the nearest standard size should be used.

25. The number and diameter of the bolts in the steam-chest, or valve-chest, cover may be found in a manner similar to that employed for the cylinder head. The bolts should not be less than $\frac{3}{4}$ inch in diameter, nor spaced farther apart than 5 inches between centers.

Let a_1 = length of valve-chest cover, in inches, between the center lines of bolts on opposite ends;

b_1 = breadth of valve-chest cover, in inches, between the center lines of bolts on opposite sides.

Then, with the other quantities the same as in Art. **24**, the load on each bolt due to screwing up is

$$L = 16{,}000\,d \qquad (1)$$

The load on each bolt due to the steam pressure is

$$l = \frac{a_1 b_1 p}{n} \qquad (2)$$

Placing the sum of the loads in formulas **1** and **2** equal to the safe stress on one bolt gives $.7854\,d^2 S_1 = 16{,}000\,d + \frac{a_1 b_1 p}{n}$, which reduces to

$$d = \frac{10{,}186}{S_1} + \sqrt{\frac{a_1\, b_1\, p}{.7854\, S_1\, n} + \left(\frac{10{,}186}{S_1}\right)^2} \qquad (3)$$

The safe tensile stress would be from about 15,000 to 17,000 pounds per square inch for wrought iron, and from 18,000 to 21,000 for steel.

26. Steam Ports and Passages.—The dimensions of steam ports, exhaust ports, and other steam passages depend on the velocity of flow of the steam. The ports and passages must be large enough to permit the steam to follow up the advancing piston without loss of pressure.

The area of cross-section of the steam and exhaust pipes may be found from the formula

$$a = \frac{A\, v}{V}$$

in which a = area of cross-section of the pipe, in square inches;
A = area of piston, in square inches;
v = piston speed, in feet per minute;
V = velocity of flow of steam in the pipe, in feet per minute.

The values of V commonly used in practical construction are 4,000 for the exhaust pipe, and 6,000 for the steam pipe, although these values are sometimes increased to as much as 6,500 and 8,000, respectively. As the pipes are circular, the internal diameters can readily be obtained from their areas of cross-section. The nearest standard size of pipe to that found in the manner just described should be used.

The area of cross-section of the steam and exhaust ports may also be found from the formula just given. The values of V used in practical construction are from 4,000 to 6,000 for the exhaust port and from 6,000 to 8,000 for the steam port, the lower values being the most common. Where the same port is used for both steam and exhaust, the area of the port must be made large enough for the exhaust. If l is the length of the port and D the diameter of the cylinder, then l is made equal to about $.7\,D$ to D, the usual value being between $.8\,D$ and $.9\,D$ for slide-valve engines, and about $.9\,D$ to D for the Corliss type.

With steam superheated about 100° F., the values of V just given may be increased from 30 to 40 per cent. in the case of the steam pipe, and if separate ports are used for steam and exhaust, the value of V for the steam port may also be increased the same amount.

The height w, Fig. 4, of the valve seat above the center line of the cylinder should be made as small as possible without interfering with the size of the steam and exhaust ports.

27. Steam-Chest Calculations.—In Fig. 5 is shown a design of a cylinder having the steam chest cast solid with it. The crank-end head in this case is a separate cast-

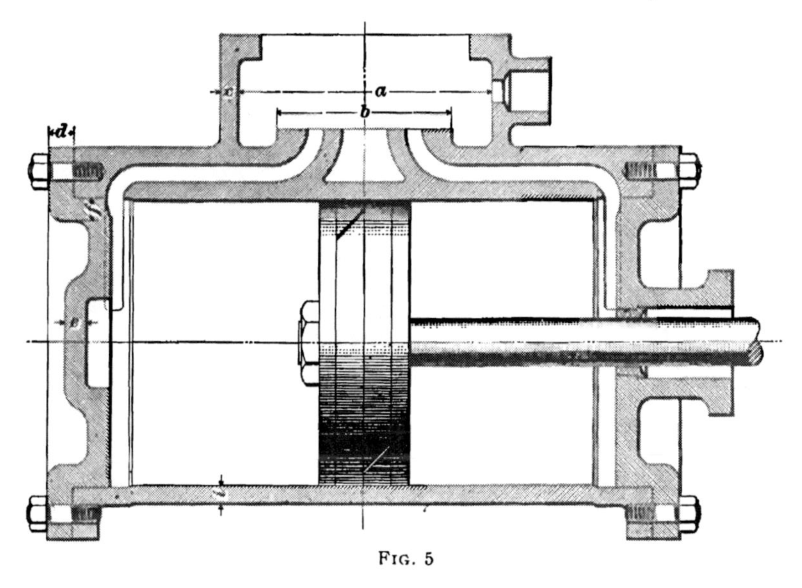

Fig. 5

ing fitted to the cylinder in the same manner as the head-end head. The heads, which are cast without ribs, are well suited for cylinders of small diameters. For larger diameters, the ribbed heads shown in Fig. 4 are better.

The following proportions apply to Fig. 5:

p = maximum steam pressure, in pounds per square inch, gauge.

D = diameter of cylinder, in inches.

$i = .0003\,p\,D + .375$ inch.

a = length of valve + travel of valve + twice the clearance between valve and steam chest at ends of valve travel.

b = valve travel + length of valve − $\frac{1}{4}$ to $\frac{1}{2}$ inch.

$c = i$.

$d = 1.5\,i$.

$e = 1.25\,i$.

$f = 1.25\,i$.

All other dimensions are to be determined by the formulas given for Fig. 4.

28. Fig. 6 illustrates a steam chest for the cylinder

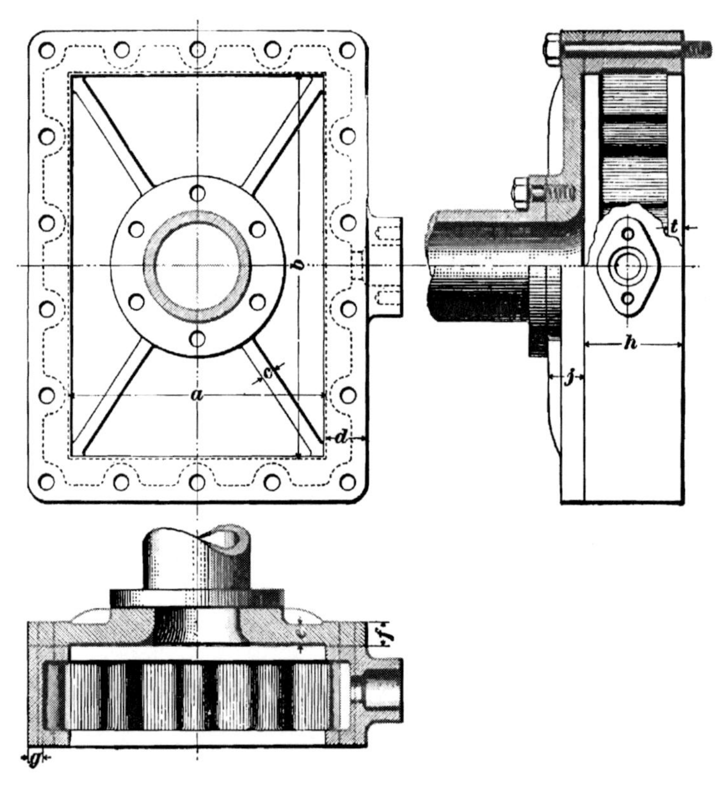

FIG. 6

shown in Fig. 4. The principal dimensions are to be determined by the following proportions, which are based on the thickness i of the cylinder walls, as found in Art. **22**, and on the travel and dimensions of the valve:

a = length of valve + travel of valve + twice the clearance between the valve and the steam chest at ends of valve travel.

b = breadth of valve + twice the clearance between one end of the valve and the steam chest.

c = $.75\,i$.

d = $2.75\,o$, where o is the nominal diameter of the steam-chest bolts.

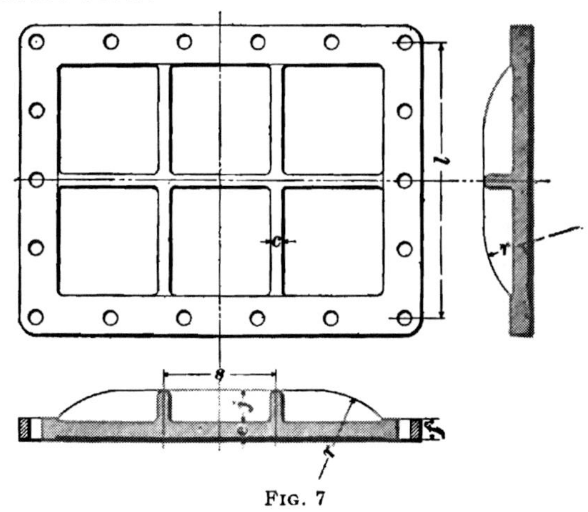

FIG. 7

e = $.05\sqrt{A'} + .125$ inch, in which A' is the area of steam-chest in square inches, obtained by multiplying the length between center lines of bolt rows by the width between center lines; but e should never be made less than i, whatever may be the result of calculation by this formula. When the steam pressure is above 100 pounds per square inch, gauge, the dimension should be changed to $e = .05\sqrt{\frac{p\,A'}{100}}$ + .125 inch.

f = $1.3\,e$.

g = i.

h = height of valve + necessary clearance.

j = $2.5\,i$. When the area of the steam-chest cover exceeds 600 square inches, the height of the ribs should be $3.5\,i$, and their number should be increased.

t = $.85\,i$.

29. Fig. 7 shows a steam-chest cover design that should be used when the steam-pipe flange is to be located on one side of the steam chest. The dimension indicated by the letters have the following values:

$e = .05\sqrt{A'} + .125$ inch for the thickness of the cover, as in Art. **28.**

$c = e.$

$f = 1.3\,e.$

$j = 2.5\,e$, at least.

$l =$ width of steam-chest cover between center lines of bolts, in inches.

$p =$ steam pressure, in pounds per square inch, gauge.

$r = 7.7\,e.$

$s =$ distance between centers of ribs, and should never exceed the distance in inches given by the formula

$$s = \sqrt{\frac{40\,e_1^2}{p}}, \qquad (1)$$

in which e_1 is the thickness of the cover in sixteenths of an inch.

To insure safety, the shorter ribs across the cover should be sufficiently strong to carry the total load and at the same time allow a small factor of safety, say about 2, without any aid from the long rib and the flat part of the cover. To obtain this result, let

$$S_1 = \frac{3}{4}\frac{p\,s\,l^2}{c\,(e+j)^2}, \qquad (2)$$

in which S_1 is the safe stress of cast iron in flexure, in pounds per square inch.

Substituting in formula **2** the values of the letters as just stated, S_1 should not exceed 15,000; if it does, either the thickness or the height of the ribs, or both, should be increased, or their distance apart should be decreased until the resulting value of S_1 does not exceed 15,000.

EXAMPLE.—Find the thickness of the cover and the thickness, height, and pitch of the ribs of the cover for a steam chest having a maximum length and width of 24 and 16 inches, respectively. The chest is subjected to a steam pressure of 160 pounds per square inch.

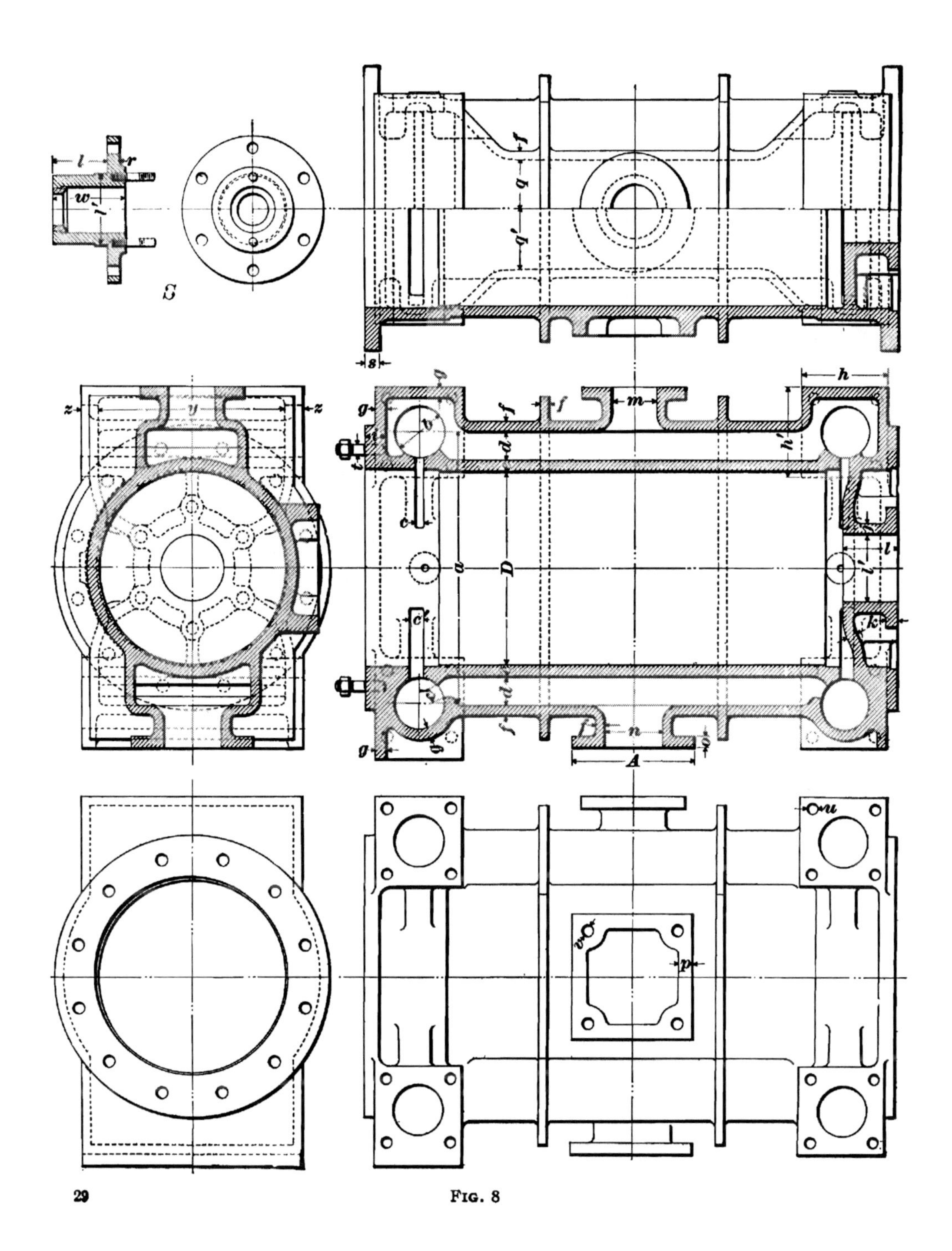

FIG. 8

SOLUTION.—The thickness of the cover is $e = .05\sqrt{24 \times 16} + .125$ in. $= 1.105$ in. Use $e = 1.125$, or $1\frac{1}{8}$, in.

The thickness of the ribs on the cover is $c = 1\frac{1}{8}$ in.

The height of the ribs above the cover is $j = 2.5 \times 1.125 = 2.81$ in. Use $j = 2\frac{7}{8}$ in.

Then, applying formula **1** for the pitch of the ribs,

$$s = \sqrt{\frac{40 \times (18)^2}{160}} = 9 \text{ in.}$$

Use two short ribs 8 in. apart and one long rib over the middle of the chest. Then, in order to be certain that the cover will be safe, apply formula **2** to find the stress in the ribs. Thus,

$$S_1 = \frac{3}{4} \times \frac{160 \times 8 \times (16)^2 \times 8}{9 \times 4 \times 4} = 15{,}360 \text{ lb.},$$

which is satisfactory.

30. Corliss-Engine Cylinder Proportions.—Corliss and other long-stroke engines, especially those having steam passages along one side and exhaust passages along the other side of the cylinder, should be designed according to the expressions given here rather than by those in Arts. **22** and **23**. Fig. 8 shows a Corliss-engine cylinder, which may be designed according to the following proportions:

D = diameter of cylinder, in inches.

$a = 1.21\,D + 2e + .125$ inch for double-ported engines, and $1.26\,D + 2e + .125$ inch for single-ported.

$b = .2\,D$ for double-ported engines, and $.25\,D$ for single-ported.

c, the width of the steam port, should be determined according to Art. **26.**

c', the width of the exhaust port, should also be determined according to Art. **26.**

$d = .17\,D$.

$e = .0005\,p\,D + .375$ inch for cylinders up to 30 inches in diameter; beyond 30 inches, e may be reduced to $e = .0004\,p\,D + .375$ inch. In the formula for e, p is the boiler pressure in pounds per square inch, gauge. The value of P in this formula, however, should never be taken less than 100.

$f = e$.

$g = e$.

$h = b + 2\,(c + g)$.

$h' = h$.

$i = 1.8e$.

$j = e$.

$k = 1.2e$.

$l = 1.7x - 1.2e + 2$ inches, where x is the diameter of piston rod, in inches.

$l' = .32D$, about.

m should equal the diameter of the steam pipe, as designed according to Art. **26**.

n should equal the diameter of the exhaust pipe, as designed according to Art. **26**.

$o = 1.25e$.

$p = 1.3e$.

$q = .25D$.

$q' = .323D$.

$r = 1.2e$.

$s = 1.5e$.

t = diameter of bolts, and should be found according to Art. **24** for cylinder-head bolts.

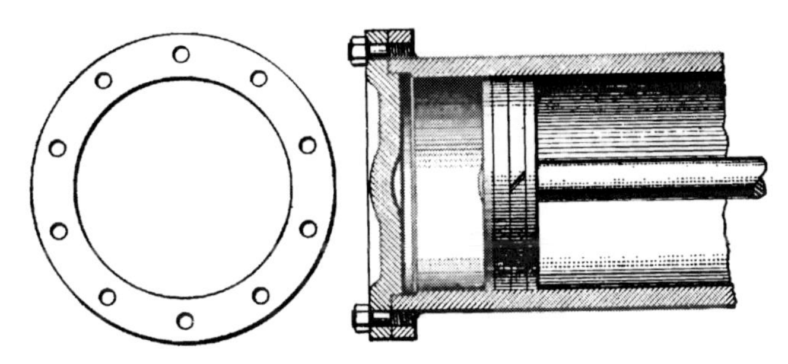

FIG. 9

u = diameter of bolts for the steam-valve chests, and may be found according to Arts. **24** and **25** for valve-chest cover bolts. It is best to use bolts of the same size on the exhaust-valve chests.

$v = 1.2e$; take nearest standard-sized bolt.

$w = 1.7x + 2.25$ inches, where x is the diameter of piston rod.

$y = .9D$ to D.

$z = 1.5e$.

Where pipes are connected, the diameter and spacing of the bolts should be made to suit the piping to be attached. In this cylinder, the stuffingbox S is a separate piece that is to be bolted to the cylinder head.

Fig. 9 shows a cylinder head suitable for cylinders of small diameter; its thickness may be made equal to the thickness of the cylinder + .25 inch.

ENGINE SHAFTS AND CRANKS

THE SHAFT

31. Diameter of Shaft.—The general dimensions of crank-shafts are computed according to the principles of machine design. The calculations are made with regard to all the principal stresses that are likely to come on the shafts. The simplest formula is that in which only the torsion of the shaft is considered, when the formula is reduced to

$$d = 68.45\sqrt[3]{\frac{H}{N S_s}}, \qquad (1)$$

in which d = diameter of shaft, in inches;
H = I. H. P. at rated load;
N = number of revolutions per minute;
S_s = safe shearing stress of material in shaft.

When there are forces tending to bend the shaft, as the weight of a flywheel, they must also be taken into consideration, and the formula then becomes more complicated.

When a series of different-sized engines of the same type are to be built, it may be assumed that they will run under about the same conditions. In such a case, it is unnecessary to use the general method of calculating the crank-shaft just given, for short empirical formulas may be deduced from the practice of the best makers.

An examination of a large number of stationary side-crank engines, all low-speed, shows the following relation between d, H and N:

$$d = K\sqrt[3]{\frac{H}{N}}, \qquad (2)$$

in which K varies from 5.66 to 7.8, the average in good practice being about 6.36.

The maximum stresses occur in the shaft in which K has the smallest value, and the minimum where K has the largest value. These stresses vary from 17,750 to 6,780 pounds per square inch, and take into consideration both torsion and bending.

32. Journal of Engine Shaft.—Certain dimensions of journals are necessary for the cool running of engine shafts.

Let l = length of main journal, in inches;
H = I. H. P. at rated load;
L = length of stroke, in inches;
d = diameter of shaft at bearing, in inches;
D = diameter of cylinder, in inches.

An examination of a large number of stationary side-crank engines, all low-speed, shows the following relations:

$$l = K_1 \frac{H}{L} + 7, \qquad (1)$$

in which K_1 varies from .86 to 2.27, the average being about 1.56.

$$d = K_2 \frac{D^2}{l}, \qquad (2)$$

in which K_2 varies from .36 to .5, the average being about .44.

The product of d and l is called the *projected area* of the journal or bearing, and the pressures per square inch of projected area found vary from 156 to 218 pounds, 178 pounds being the average. Some authorities recommend values as high as 450 pounds per square inch, but an investigation of a large number of engines in use shows the values just given. The diameter of the shaft must be calculated for both strength and cool running, and the larger values of d found should be used.

33. An examination of stationary center-crank engines, all high-speed, shows the following relations:

$$d = K_3 \sqrt{\frac{H}{N}}, \qquad (1)$$

in which the value of K varies from 5.98 to 8.76, the average value in good practice being 7.56,

$$l = 1.1\frac{H}{L} + 4 \qquad (2)$$

$$d = .4\frac{D^2}{l} \qquad (3)$$

Formula **3** should give the diameter of the main journal for cool running in high-speed center-crank engines. This corresponds to a pressure of about 100 pounds per square inch of projected area of the bearing surface, due to the steam pressure alone, assuming that the steam pressure in the cylinder is 100 pounds per square inch and that the pressure on the crankpin is transmitted equally to the two bearings. The diameter of the shaft must be calculated for both strength and cool running, and the larger diameter found must be taken, just as with low-speed engines.

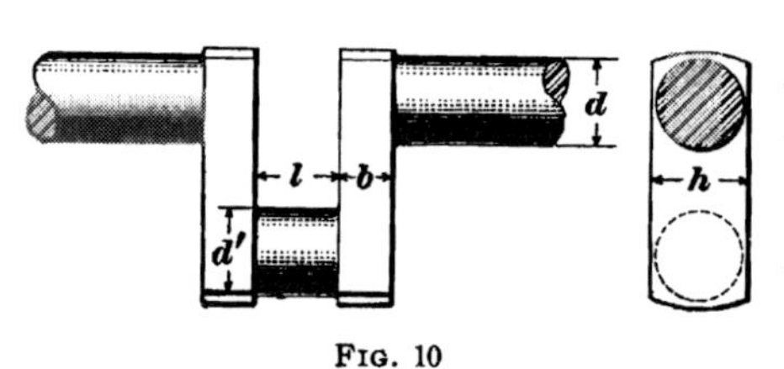

FIG. 10

A solid-forged double crank is shown in Fig. 10, the crank, or cranks, being forged in the main shaft. When the ratio $\frac{H}{N}$ is not less than **1** and $K = 4.55$, formula **1** may be used to find the diameter of the shaft at the journals. The diameter d' and length l of the crankpin will be considered in the next article, and the only other dimensions to be found are b and h.

Usually, $b = .6\,d$ to $.8\,d$,

and $$h = 1.53\sqrt{\frac{D^2 p r}{b S_1}}, \qquad (4)$$

in which p = maximum unbalanced pressure on the piston, in pounds per square inch, which should not be taken less than 113;

S_1 = safe tensile stress in the material, in pounds per square inch;

D = diameter of steam cylinder, in inches;

r = length of crank, in inches.

The other quantities are the same as already given.

THE CRANKPIN

34. Crankpins for Overhung Crank.—A crankpin should be designed for strength, rigidity, and cool running. The case of the **overhung crank,** that is, a crank such as is shown in Fig. 11, will be considered first. Some variations in crankpin proportions will be noted in practice, and as a guide as to choice among these variations, it may be stated that in all classes of engines there is a growing tendency toward larger diameters of crankpins and smaller ratios of

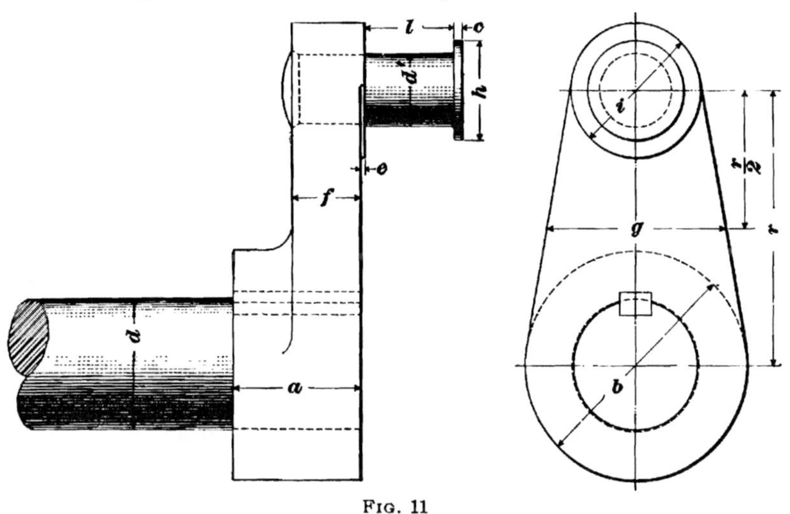

FIG. 11

length of pin to diameter; this is especially true of the center-crank engine.

Let d' = diameter of crankpin, in inches;
l = length of crankpin, in inches;
p = maximum unbalanced steam pressure, which should not be taken less than 113;
p_1 = pressure on crankpin in pounds per square inch of projected area;
k = deflection of outer end of overhung crankpin, in inches;
H = I. H. P. at rated load;
D = diameter of cylinder, in inches;
L = length of stroke, in inches.

For cool running, the length of a journal, such as a crank-pin, should theoretically be given by the following formula, in which K is a constant:

$$l = \frac{K H}{L} \qquad (1)$$

An examination of the proportions used in practice shows that the length is given more closely by an expression of the form

$$l = K\frac{H}{L} + K_1 \qquad (2)$$

An examination of side-crank, low-speed stationary engines shows that the value of K_1 is equal to 2 in all cases, and that in good practice K varies from .345 to .655.

For cool running, the following formula applies in all cases:

$$d' = .7854\frac{D^2 p}{p_1 l} \qquad (3)$$

Some authorities place the value of p_1 between 800 and 900 for low-speed, side-crank stationary engines, but an examination of actual engines shows that this value in good practice runs from 873 to 1,570.

35. For strength, the following formula for the diameter may be used in all cases:

$$d' = \sqrt[3]{\frac{4 p D^2 l}{S_1}}, \qquad (1)$$

in which S_1 is the safe tensile stress. With the low-speed, side-crank stationary engine, S_1, in good practice, varies from 3,200 to 12,500, the average value being about 7,000.

When p and S_1 are assigned numerical values, formula **1** may be reduced to the expression

$$d' = K_2\sqrt[3]{D^2 l}, \qquad (2)$$

in which K_2 represents the factor $\sqrt[3]{\frac{4p}{S_1}}$.

An examination of low-speed, side-crank stationary engines shows that in good practice K_2 varies from .32 to .5, with .384 as the average value.

36. For stiffness, the following formula gives the value of d':

$$d' = .016\sqrt[4]{\frac{p D^2 l^3}{k}} \qquad (1)$$

In low-speed, side-crank stationary engines, the deflection k, in good practice, varies from .0004 to .005, the average value being about .001.

Where p and k have numerical values, the formula just given reduces to the form

$$d' = K\sqrt[4]{D^2 l^3} + K_1 \qquad (2)$$

An examination of low-speed, side-crank stationary engines shows that, in good practice, K_1 varies from .6 to 1, with an average value of .81, and that K varies from .185 to .353, with an average value of .28.

For cool running, for strength, and for stiffness, the diameter of the pin should be found independently, and the largest value should be used. Generally, this value will be given by the calculations for stiffness, but this should not be taken for granted, and all three calculations should be made.

37. Crankpin for Center Crank.—Center crankpins are found to have very different dimensions than overhung pins. For cool running, instead of using the theoretical expression, the length of the pin is found to be given more closely in practice by the formula

$$l = K\frac{H}{L} + K_1 \qquad (1)$$

An examination of center-crank, high-speed stationary engines shows that when $K = .35$ and $K_1 = 2.2$ inches, formula **1** gives the average value of l in good practice; but many pins show wide departures from average practice.

For cool running, the diameter for all cases is found by the formula

$$d' = .7854\frac{D^2 p}{p_1 l} \qquad (2)$$

With high-speed, center-crank stationary engines, p_1 varies from 225 to 1,400, the average being about 450. With marine engines, p_1 varies from about 400 to about 600.

Where p and p_1 have definite values, formula **2** reduces to

$$d = K\frac{D^2}{l}, \qquad (3)$$

in which K represents $\frac{.7854\,p}{p_1}$.

An examination of high-speed, center-crank stationary engines, shows that K varies from .089 to .599, with an average value of .28 in good practice.

A purely theoretical consideration of the strength and stiffness of the center-crank pin would indicate that it might be of smaller diameter than the shaft, but in practice this is found to be unsatisfactory. The diameters of center-crank pins in actual engines are always about the same as those of the main journals, being sometimes a little less, but greater as a rule. It may therefore be considered an empirical rule, sufficiently established by practice, that a center-crank pin should be of about the same diameter as the main journals—better greater than smaller. If the diameter necessary for cool running is greater than the diameter of the main journals, this greater diameter should be taken.

CRANK AND COUNTERBALANCE

38. Fig. 11 shows a style of crank much used on low-speed engines, such as the Corliss type. Let S be the maximum stress on the material, in pounds per square inch; and r the length of the crank, in inches. The other symbols represent the same quantities as given in Arts. **31** to **37.** The dimensions are to be computed in the following manner:

For d, use the method explained in Art. **31.**

a = .75 to 1 times the diameter of the shaft. An effort should be made to keep down the value of a, as by so doing the bending moment on the main bearing is decreased.

To obtain sufficient strength opposite the key shown in Fig. 11, b should be found by the following formula:

$$b = .75\,d + .5\sqrt{\frac{d^2}{4} + \frac{1.57\,D^2 p\,r}{a\,S}} \qquad (1)$$

S may be taken as 9,000 for wrought iron and 11,000 for steel; p is the maximum unbalanced steam pressure, but should never be taken less than 113 pounds per square inch.

After finding b by means of the formula just given, the stress in the boss at right angles to the crank should be found by the formula

$$S_4 = \frac{4.71\, D^2 p\, r\, b}{a(b^3 - d^3)} \qquad (2)$$

If S_4 as thus calculated exceeds the safe bending stress for the material, b should be increased until S_4 is sufficiently reduced.

$c = .045\, d + .0625$ inch.

d' should be calculated as explained in Arts. **34** to **36.**

$f = .375 g$, as a trial value. If desirable, this may be increased later.

g is found by drawing lines tangent to the circles b and i, and taking the distance between these lines at a point midway between the main shaft and crankpin centers. The stress at g should then be found by the formula

$$S_4 = \frac{2.36\, D^2 p\, r}{g^2 f} \qquad (3)$$

As before, p should not be taken less than 113. If the value of S_4 as thus found is beyond the safe bending stress of the material, f should be increased. If f cannot be increased enough to bring the stress down to a safe value, b and i should be increased.

$h = 1.35\, d'$.

As a trial value, i may be made $2\, d'$; after f and g are found, i may be calculated by the formula

$$i = \frac{.7854\, D^2 p}{f S_1} + d' \qquad (4)$$

The remarks in regard to the use of p and S_4 given in connection with formula **3** apply equally as well to p and S_1 in this formula. S_1 is the safe tensile stress in the material.

l should be calculated the same as in Art. **34.**

39. Modern high-speed engines are often counterweighted to reduce vibrations in running, and the crank in such cases usually takes the form of a disk, as shown in Fig. 12. The disk is hollowed out as shown, but a portion of the material is left in the side opposite the crankpin so as to form the counterweight, which, by its centrifugal force, counteracts the centrifugal force resulting from the motion of the reciprocating and other rotating parts of the engine. As here shown, the counterweight may be made as a separate part from the disk proper, it being attached only to the hub *a*, with which it is cast in one piece; it is thus made to allow for expansion and contraction in the larger crank-disks. The

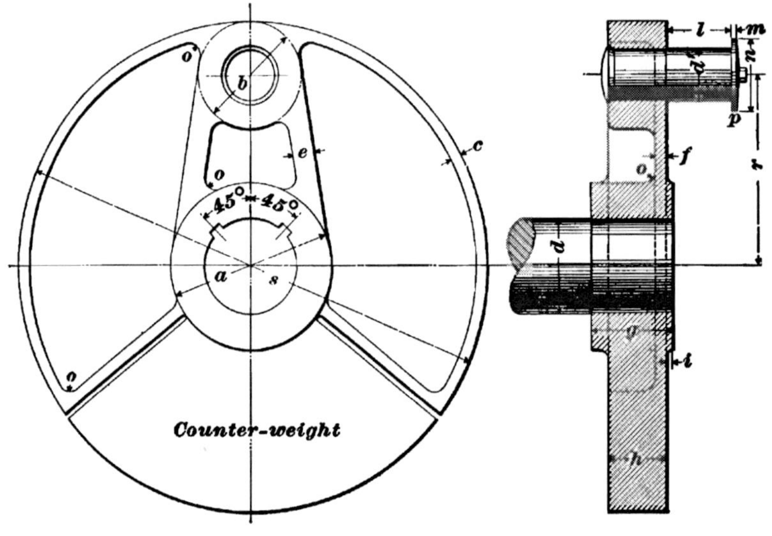

FIG. 12

width of the split is about $\frac{3}{4}$ inch for engines of 48-inch stroke or less, and 1 inch for all larger sizes.

In Fig. 12, p is a plate held to the end of the crankpin by a tap bolt. The radius of all fillets except that of the boss i is o. Assume that s is the diameter of the crank-disk in inches; r, the length of the crank-arm, in inches; D, the diameter of the cylinder, in inches; p, the maximum unbalanced pressure on the piston, in pounds per square inch, which should never be taken as less than 113; S, the maximum stress on the material, in pounds per square inch; and k, the deflection of the outer end of the pin, in inches. The significance of all other symbols used in this article is shown in Fig. 12. The different dimensions should be determined from the following rules:

a should be determined by formula **1**, Art. **38**, for finding *b* of Fig. 11, but the calculation of the stress in the section of the hub at right angles to the center line of the crank, as found in formula **2**, Art. **38**, for Fig. 11, may be omitted.

b should be calculated by formula **4**, Art. **38**, for finding *i* of Fig. 11.

$$c = \frac{r}{24} + .5 \text{ inch}$$

d' should be calculated for cool running by the formulas of Art. **34**. If the rotative speed is above 125 revolutions per minute, the formulas of Art. **37** for high-speed center-crank engines should be used. If the rotative speed is not greater than 125 revolutions per minute, the formulas of Art. **34** for low-speed side-crank engines should be used.

For strength, use formula **1**, Art. **35**. S_1 may be taken as 9,000 for wrought iron, and 11,000 for steel.

For stiffness, use formula **1**, Art. **36**, taking the deflection *k* as .001, as in Art. **34**.

d' should be calculated separately for cool running, strength, and stiffness, and the greatest value found should be taken.

$e = 2c$.

$f = c$.

g should be somewhat greater than *h*, and is usually about .875 *d*.

h = thickness of counterweight, to be calculated so as to make the counterweight as heavy as necessary.

i should be given such a value that the connecting-rod will have about $\frac{1}{8}$-inch clearance.

l should be calculated by the formulas given in Arts. **34** or **37**. If the rotative speed is above 125 revolutions per minute, the formulas of Art. **37** for high-speed center-crank engines should be used. If the rotative speed is 125 revolutions per minute or less, the formulas of Art. **34**, for low-speed side-crank engines should be used.

$m = .045\,d + .0625$ inch.

$n = 1.35\,d'$.

$o = c$.

$s = 2\,r + b$, for the crank-disk proper and generally for the counterweight also, but, if necessary, the value of s for the counterweight may be increased or diminished in order to get the proper amount of counterweight. This method is objectionable, however, and should be avoided if possible.

40. Counterbalancing the Crank.—In order to find the thickness of the counterweight used on a crank like the one shown in Fig. 12, it is first necessary to find the weight.

Let W_1 = weight of counterweight, in pounds;

W_2 = weight of crank, outside of hub around main shaft, in pounds;

W_3 = weight, in pounds, of the reciprocating parts; that is, weight of piston + weight of piston rod + weight of crosshead + one-half the weight of connecting-rod;

X = distance of center of gravity of counterweight from center of shaft;

Y = angle between the two radial sides of counterweight, this angle being always equally divided by the center line of the crank.

Then, the distance X is found by the formula

$$X = \frac{38.2\,(s^3 - a^3)\sin\frac{Y}{2}}{(s^2 - a^2)\,Y} \qquad (1)$$

When the angle Y is known in degrees, then the formula may be written

$$X = K\frac{s^3 - a^3}{s^2 - a^2} \qquad (2)$$

Thus, when the angle $Y = 60°$, $K = .32$; when $Y = 90°$, $K = .3$; when $Y = 120°$, $K = .276$; when $Y = 180°$, $K = .21$.

To counterbalance so as to avoid knocks when the crank is at right angles to the center line of the engine, the weight of the counterbalance is found by the formula

$$W_1 = \frac{W_2\,r}{X} \qquad (3)$$

To counterbalance so as to avoid knocks when the crank passes over dead center, the counterweight is found by the formula

$$W_1 = \frac{(W_2 + W_3) r}{X} \qquad (4)$$

For the greatest possible smoothness of running throughout the revolution, use the formula

$$W_1 = K \frac{(W_2 + W_3) r}{X}, \qquad (5)$$

in which K has a value of from .67 to .75, the smaller value being used where it is desired to have absence of vibration when the crank is on the quarter, that is, at right angles to the center line of the engine, and the larger value where absence of vibration is most desirable when the crank is at dead center.

In general, the counterbalancing of vertical engines should tend toward the greatest possible smoothness of running, with the crank on the quarter; and that of horizontal engines, with the crank on dead center.

As the counterweight is usually made of cast iron, the thickness h for this material may be found by the following formula, in which Y is in degrees, s and a in inches, W_1 in pounds, and the weight of cast iron is taken as .26 pound per cubic inch:

$$h = 1{,}763 \frac{W_1 (s^2 - a^2)}{Y} \qquad (6)$$

THE PISTON

PISTON BODY

41. Hollow Pistons.—Engine pistons are made in a great variety of forms. For small engines, that is, for cylinder diameters less than 8 or 10 inches, the piston is often a solid disk of cast iron or steel. A form of **hollow piston** that is much used for small engines is shown in Fig. 13, and consists simply of a hollow circular disk of cast iron. The packing rings s, s are made of cast iron and are split and sprung into place; their elasticity causes them to press against the cylinder walls and thus prevents the leakage of steam. The following proportions will give dimensions suitable for this piston:

D = diameter of cylinder, in inches.

For high-speed engines, $a = .3\,D$, as minimum value; $.46\,D$, as average value; and $.6\,D$, as maximum value.

For low-speed engines, $a = .25\,D$, as minimum value; $.32\,D$, as average value; and $.45\,D$, as maximum value.

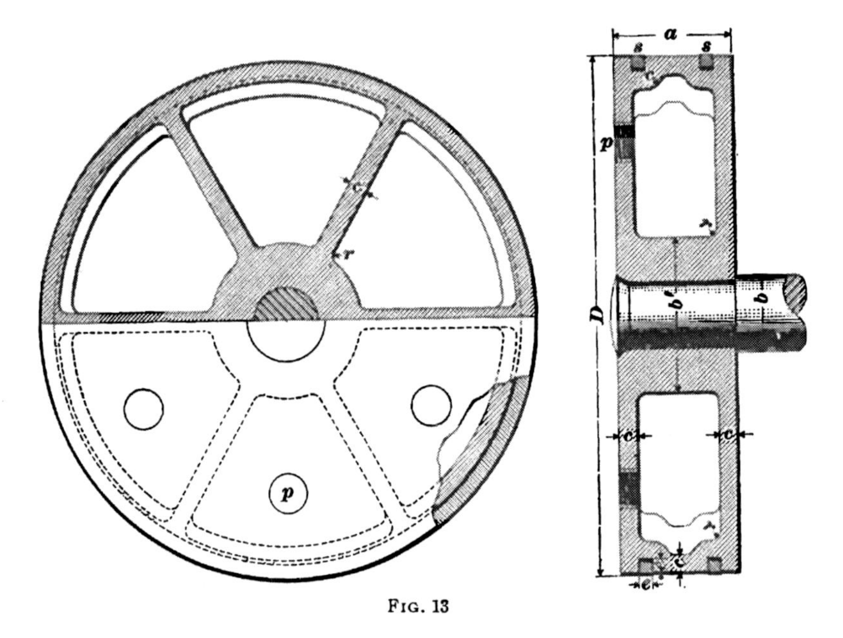

FIG. 13

Both the stiffness and the strength of the piston are greatly increased by increasing the value of a; on the other hand, the weight is increased, and, in horizontal engines, the friction between the piston and the cylinder is also increased.

b = diameter of piston rod.

$b' = 2\,b$.

$c = .18\sqrt{2\,D}$.

$e = .75\,c$.

$r = .5\,c$.

p is a core plug of suitable diameter.

Number of ribs $= .08\,(D + 34)$.

42. Built-Up Pistons.—The piston shown in Fig. 14 is made in two parts; the main part A is called the *spider*, to which is bolted the *follower plate* B. The spider is cast hollow, with radiating arms and lugs for the follower bolts i. A split cast-iron *bull ring* b is placed around the spider, and is supported by steel springs e, which are in turn supported by brass studs f. The bull ring forms a support for the packing rings s, s.

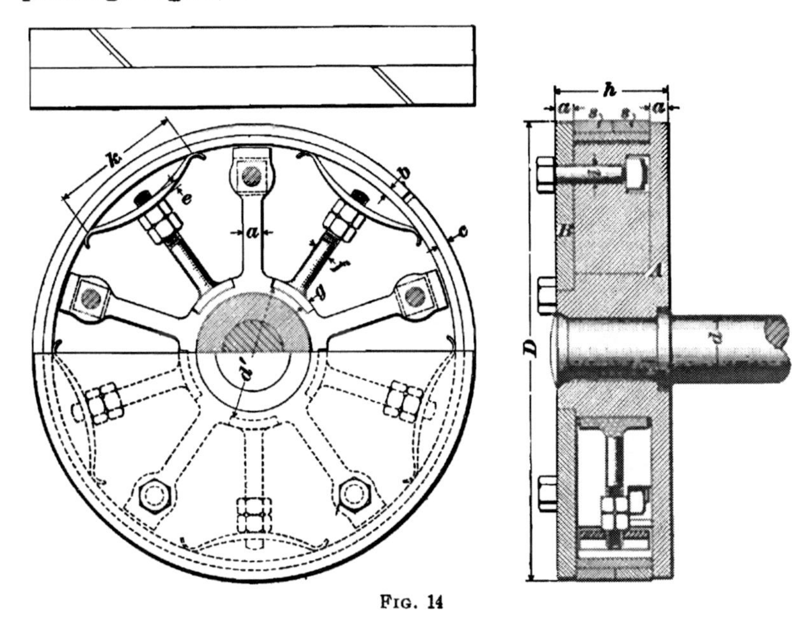

Fig. 14

The dimensions of this piston are given by the following proportions:

D = diameter of cylinder, in inches.

$a = .18\sqrt{2D}$.

$b = .45a$.

$c = .65a$.

d = diameter of piston rod.

$d' = 2d$.

$$e = \frac{.06D}{n}.$$

$f = a$.

$g = .5f$.

h should be calculated by the rules given for a in Art. **41.** In the built-up piston, h should tend rather toward the maximum of practice, as a follower plate does not contribute nearly so much to the strength of the piston as does a solid back cast on the piston. The strength of the built-up piston is derived chiefly from the radiating arms a of the spider, the strength of which is greatly increased by increasing their depth.

$i = .1\left(\frac{D}{50}\sqrt{P} + 1\right) + .25$ inch, where p is the maximum unbalanced pressure on the piston, in pounds per square inch, which should never be taken as less than 113.

$k = 1.4\frac{D}{n}$.

n = number of ribs $= .08(D + 34)$.

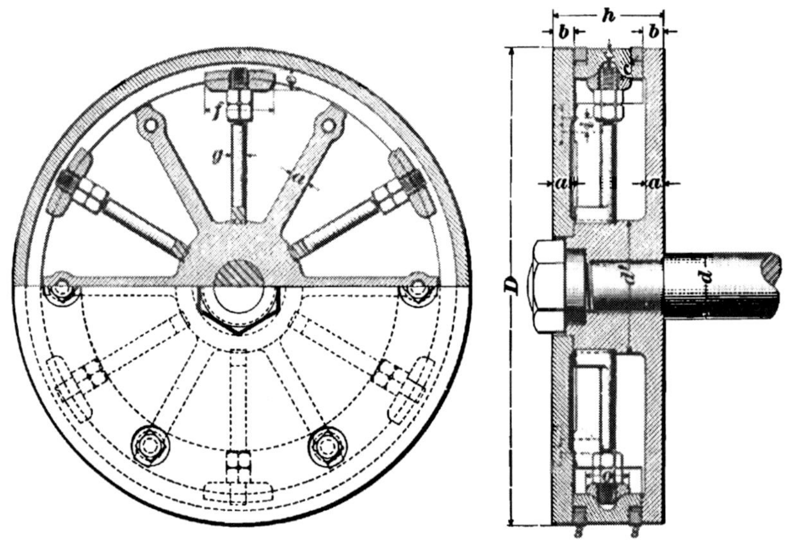

Fig. 15

43. Another form of built-up piston is shown in Fig. 15. The proportions to be used for this piston are:

D = diameter of cylinder, in inches.

$a = .18\sqrt{2D}$.

$b = 1.25a$.

$c = .75a$.

d = diameter of piston rod.

$d' = 2d$.

$e = 1.25a$.

$f = 3.75a$.

$g = a$.

h should be calculated by the rules given for a in Art. **41.** In the built-up piston, the depth h should tend rather toward the maximum of practice, as the follower plate does not contribute nearly so much to the strength of the piston as does a solid back cast in one piece with the piston.

$i = a - .125$ inch, but should never be reduced below .75 inch.

$o = 2.5a$.

n = number of ribs $= .08(D + 34)$.

44. Solid Pistons.—A form of piston much used in locomotive practice is shown in Fig. 16. A piston of this

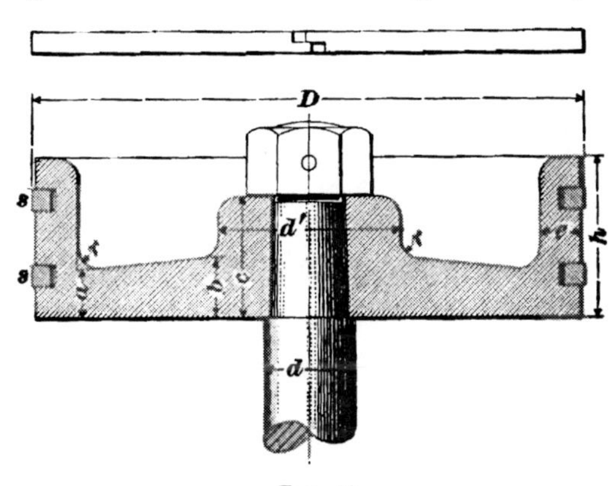

Fig. 16

kind may be made of cast iron, but it is usually a steel casting. Suitable proportions for a cast-iron piston are:

D = diameter of cylinder, in inches.

$c = .15D + 1.125$ inches.

d = diameter of piston rod.

$d' = 2d$.

$e = .27\sqrt{2D}$.

$h = .2D + 1.5$ inches.

$r = .5e$.

The thickness of the body, or disk, of the piston at any point between the boss and the flange should be calculated by the following formula:

$$T = \sqrt{\frac{3p\left(\frac{D^2}{4} - \frac{x^2}{3}\right)}{S}},$$

in which T = thickness, in inches, of piston disk;
p = maximum unbalanced pressure on piston, in pounds per square inch;
D = diameter of cylinder, in inches;
x = distance, in inches, from center of piston to point at which thickness is to be found;
S = maximum stress in material at point, which should not be taken as more than 3,000 pounds per square inch for cast iron.

45. If the piston is to be a steel casting, the following proportions should be used, all other dimensions being the same as for a cast-iron piston:

$d' = 1.75\,d.$

$e = .23\sqrt{2\,D}.$

The thickness of the body, or disk, of the piston should be calculated by the same formula as for cast iron, but S may be given a value as high as 5,500 pounds per square inch.

In either case, it is preferable to make the front of the piston flat, as shown in Fig. 16. The thickness at the point a just at the inner edge of the fillet between the disk and the flange and the thickness b just at the outer edge of the fillet between the boss and the disk should be calculated by the formula just given. These thicknesses should then be laid off on the drawing from the flat face of the piston. A straight line drawn between the two points of the back thus determined will give the outline of the rest of the back of the disk.

Since the piston is solid, the rings must be cut and sprung into place. In order to make the clearance small, the form of the cylinder head is made to conform to the piston.

46. Marine Pistons.—Fig. 17 shows a piston made of a steel casting designed for the high-pressure cylinder of a marine engine. The dotted lines show how the cylinder head is made in order to fit closely around the piston. A conical form is given the piston in order to increase its strength.

The following proportions will give suitable dimensions for this piston for cylinders from 7 to 50 inches in diameter:

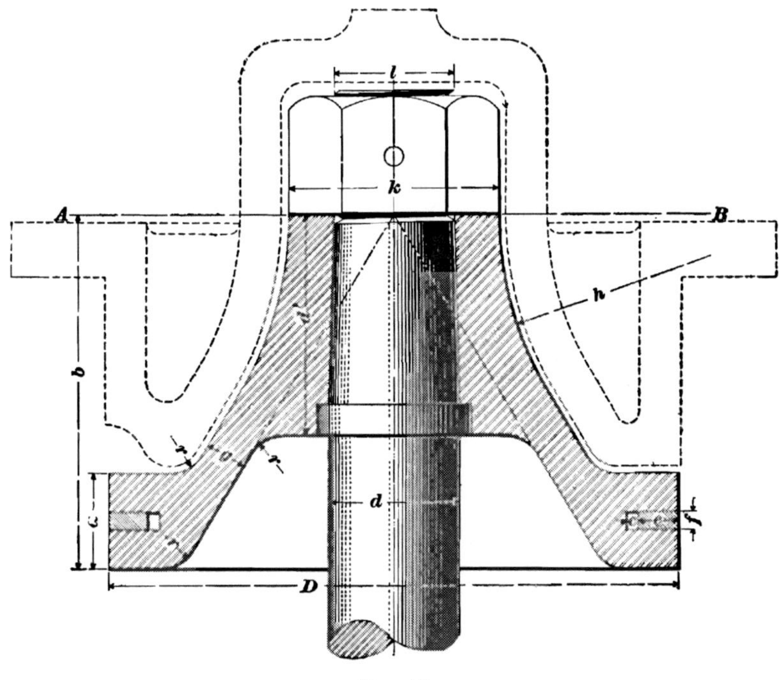

FIG. 17

D = diameter of high-pressure cylinder, in inches.

$a = .1\,D + 1.5$ inches.

$b = \sqrt{20\,D} - 7.5$ inches.

$c = .1\,\sqrt{D}$.

d = diameter of piston rod.

$d' = 1.63\,d$. This expression gives values of d', which are amply large.

The use of a large boss has its own dangers. In casting, the body of metal produced in the center of the piston may remain fluid after the rest of the piston has set, and then, on cooling, may produce permanent stresses in the piston or even tear it apart. The same is true wherever a sudden great variation of thickness is introduced in a casting; therefore, this condition should always be avoided if possible. The designer must depend on his knowledge of foundry practice to avoid such dangers; and whenever he is in doubt it is advisable to consult the foundry foreman.

47. If the boss of the piston in Fig. 17 should appear to be too heavy to cast well, its depth may be reduced; in good practice, d' is often made as small as $.09\,D$. When d' is very much reduced, the shearing stress on the boss should be calculated, and should not exceed 5,000 pounds per square inch.

$e = .225\,\sqrt{D} + .625$ inch.

$f = .1\,\sqrt{D} + .25$ inch.

$g = .07\,D$.

h is found by trial, with the center on the line AB.

$k = 1.73\,l + .1875$ inch.

l = diameter of threaded end of piston rod.

$r = \dfrac{(a - f)}{2}$.

48. In Fig. 18 (a) is shown a piston suitable for the intermediate and low-pressure cylinders of the engine for which the piston shown in Fig. 17 is designed.

To compute the dimensions for this piston, make the dimensions a, b, c, d, d', e, f, k, l, and r, the same as for the high-pressure piston in Fig. 17. If the engine is compound, let D' represent the diameter of the low-pressure cylinder in inches and make $s = .02\,D'$. If the engine is of the triple-expansion type, let D' represent the diameter of the intermediate cylinder and compute s as before, using the value found for both intermediate and low-pressure pistons. Make the remaining dimensions as follows:

$m = 1.8\,d$.

n is to be found by trial, with the center on the line representing the bottom edge of the piston.

o is to be found by trial, with the center on the line AB.

$t = .2\,\sqrt{2\,D}$ and $t' = .275\,\sqrt{2\,D}$, in which D is the diameter of the cylinder for which the piston is made.

The formulas given for the thickness of the body of the piston in both cases make the piston decidedly heavy. Where lightness is desired, as is often the case with a conical piston, when it is a steel casting, the thickness of the piston near the flange, as at t, Fig. 18 (a), may be made from $.0015\,D\sqrt{p}$ in large engines to $.003\,D\sqrt{p}$ in small engines, but should never be less than .625 inch, p being the maximum unbalanced steam pressure on the piston in pounds per square inch. The thickness of the body of the

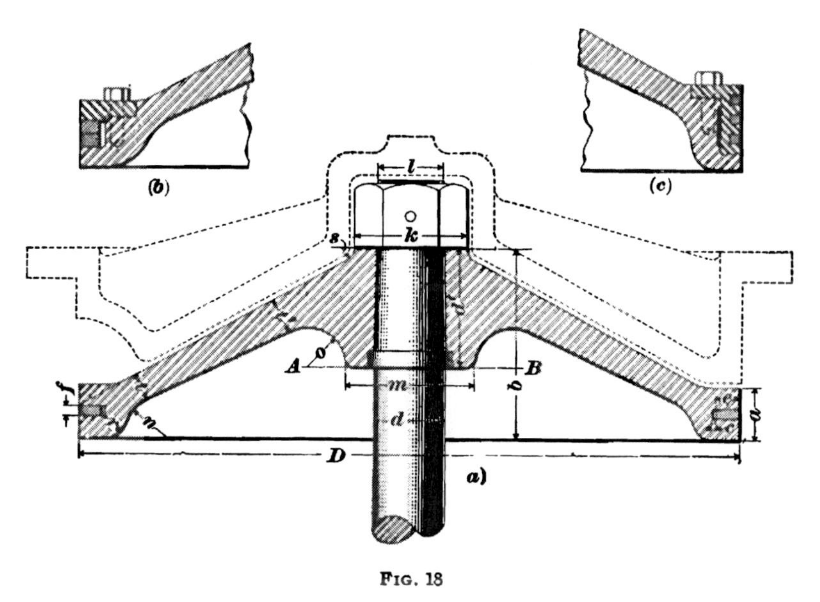

FIG. 18

piston near the boss, as t', may be made 1.75 times the thickness of the body near the flange.

49. If desired, the stress per square inch on the body of the piston near the hub, which will be the maximum stress existing in the piston, may be calculated by the formula:

$$S = \frac{p \operatorname{cosec}^2 a\,(2\,D^3 - 3\,k\,D^2 + k^3)}{12\,t'\,(D - k)^2 \operatorname{cosec} a \cot a + 4\,t'^2\,(k - t' \cos a)},$$

in which D = diameter of cylinder, in inches;
S = stress in material, in pounds per square inch;
k = diameter of piston boss;

p = maximum unbalanced steam pressure on piston, in pounds per square inch, which should not be less than 113;
a = angle between outside of conical body of piston and axis of piston rod;
t' = thickness of body of piston near the boss.

If S, as calculated by the foregoing formula, has an excessive value, the dimensions of the piston should be modified.

In Fig. 18 (*a*) and (*b*) are shown methods of attaching the packing rings to large pistons so that the rings may be removed for inspection or repairs without taking the piston out of the cylinder. This is a very important advantage in many cases, especially for marine work, where the pistons are often very heavy and facilities for handling them are poor.

PISTON PACKING

50. It is, of course, impossible to turn the piston to exactly fit the cylinder at all temperatures; therefore, the piston is made slightly smaller than the cylinder bore, and some form of packing is used to prevent the steam from leaking through between the piston and cylinder walls.

The simplest and about the best form of packing, particularly for small pistons, is the cast-iron ring shown in cross-section at s, s, Figs. 13 and 15. These rings are generally of uniform thickness. Many makers, however, prefer to make the thickness where the rings are cut about half the thickness at the opposite side.

The proportions used for the spring packing ring shown in Figs. 13 and 15 are as follows:

Thickness and depth of rings the same and equal to

$$.135\sqrt{2D} - .14 \text{ inch},$$

where D is the diameter of the cylinder.

In Fig. 19 (*a*) and (*c*), the packings for the pistons shown in Figs. 17 and 18 are shown in detail. In addition to the dimensions given in connection with the pistons, the following proportions are to be used:

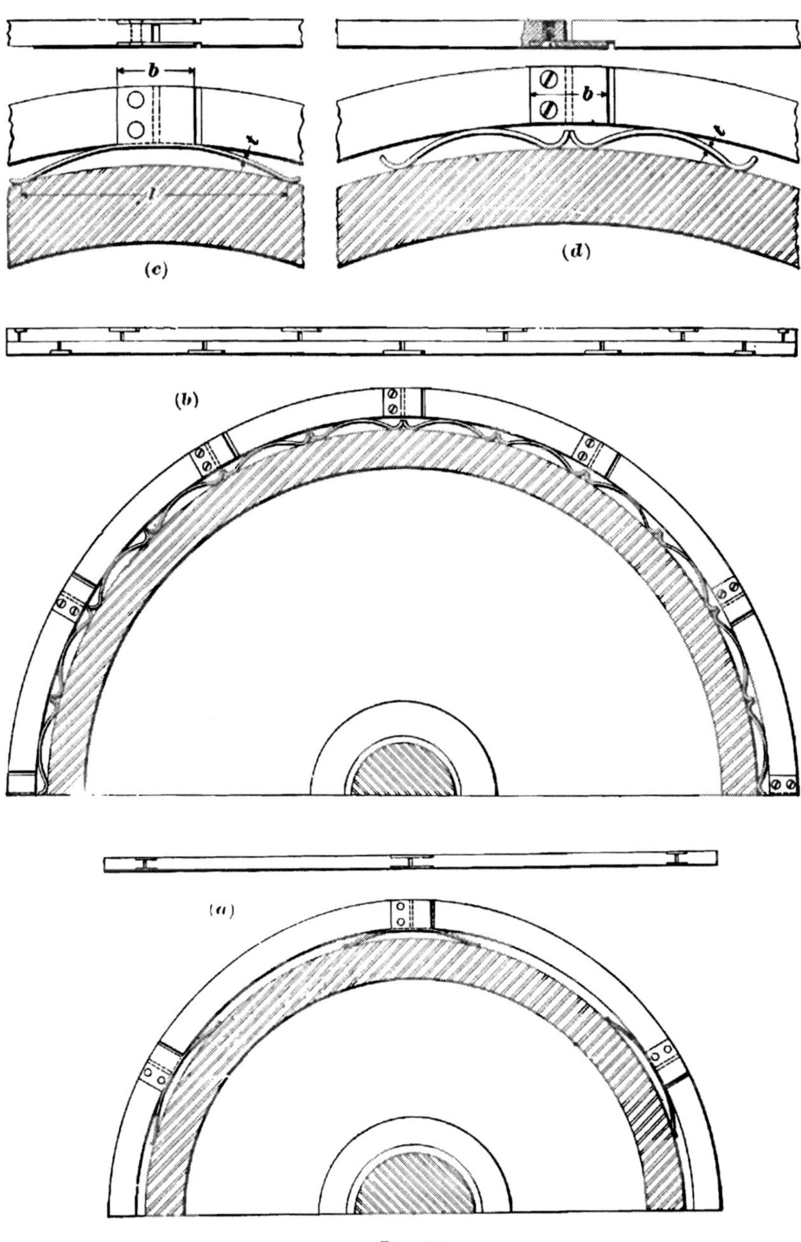

FIG. 19

b = 3 inches.
l = 9 inches.
t = .0625 inch.

Details of the packing shown in section in Fig. 18 (b) are shown in Fig. 19 (b) and (d). The proportions applying here are:

b = 3 inches.
t = .09375 inch.

The length of the segments should be about 15 inches; two springs are placed behind each segment. The packing rings shown in Fig. 19 are usually made of cast iron and the springs of steel.

51. Fig. 20 shows **Tripp's patent piston packing.** The rings s, s are made of cast iron, being split so as to

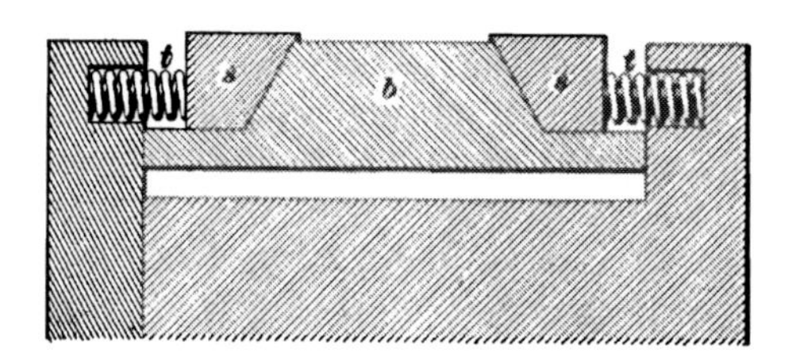

FIG. 20

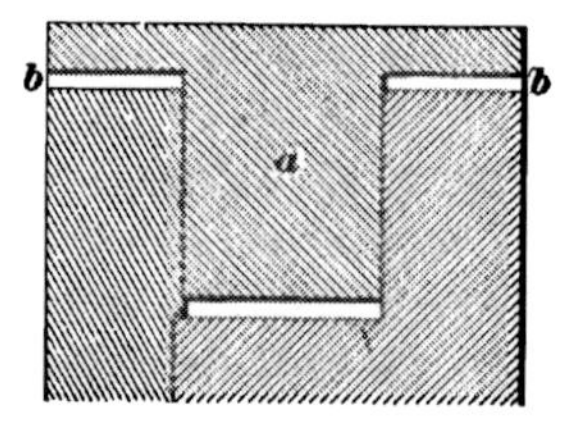

FIG. 21

spring outwards against the cylinder walls. They are supported by an adjustable ring b made with conical surfaces against which the packing rings bear. The pressure of the steam against the packing ring forces it against this conical surface, thus tending to open out the ring and make it press against the cylinder. The spiral springs t, t are for the purpose of holding the packing rings in place when they are not acted on by steam pressure.

Fig. 21 shows a style of ring packing much used for piston valves. At a is a split or sectional cast-iron ring, which is forced out against the walls of the cylindrical valve seat by the pressure of the steam in the spaces b, b between the overhanging parts of the ring and the main part of the piston or valve.

PISTON ROD

52. Piston-Rod Calculations.—The piston rod of the ordinary double-acting steam engine is subjected to alternate tension and compression of substantially equal magnitude. With the usual proportions of length to diameter, the rod should be treated as a long column, or strut, under the action of compression. As a column under compression, the rod should be considered as fixed transversely at one end and free at the other; this is to be distinguished from a column fixed at one end and guided at the other, which is eight times as strong. The formula used for such cases is based on Euler's column formula. The diameter of the piston rod is therefore given by the formula:

$$d = 2\sqrt[4]{\frac{4 f p D^2 L^2}{10 E}}, \qquad (1)$$

in which d = diameter of piston rod, in inches;

f = factor of safety, which in low-speed engines has an average value of about 10, and in high-speed engines, an average value of about 20, in good practice;

p = maximum unbalanced steam pressure on piston, in pounds per square inch, taken not less than 113;

D = diameter of cylinder, in inches;

L = length of piston rod from piston to crosshead, in inches;

E = modulus of elasticity of material, which may be taken as 25,000,000 for wrought iron and 30,000,000 for steel.

With a given material and a given steam pressure, such that E, f, and p may be taken as constant, formula **1** reduces to the form

$$d = K\sqrt{D L}, \qquad (2)$$

in which K represents $2\sqrt[4]{\frac{4 f p}{10 E}}$.

An examination of a large number of steel piston rods of stationary engines designed for a steam pressure of 100 pounds, gauge, shows the following values of K: In good practice, for high-speed engines, K varies from .119 to .177, with an average value of .145, and for low-speed engines, K varies from .098 to .136, with an average value of .112.

Formula **1** is rational, and may be used in all cases, while formula **2** is more convenient, and may be used directly in calculations for ordinary stationary engines. For other classes of work, formula **2** should not be used, except to check the results obtained by the first formula.

EXAMPLE.—Calculate the diameter of a steel piston rod 21 inches long, the diameter of the cylinder being 13 inches, the steam pressure 90 pounds, and the speed 120 revolutions per minute.

SOLUTION.—Though the expected steam pressure is only 90 lb., the rod should be calculated for a maximum unbalanced pressure of 113 lb. per sq. in. As the speed is 120, the engine is low-speed, and f may be taken as 10. Then, substituting in formula **1**,

$$d = 2\sqrt[4]{\frac{4 \times 10 \times 113 \times 13 \times 13 \times 21 \times 21}{10 \times 30{,}000{,}000}} = 2.059, \text{ say } 2\tfrac{1}{16}, \text{ in.}$$

Check the result by using formula **2**, with $K = .112$. Then, by substituting,

$$d = .112\sqrt{13 \times 21} = 1.85, \text{ say } 1\tfrac{7}{8}, \text{ in}$$

As this result is somewhat smaller than the value found by formula **1**, the value of K might be taken as .136 and again substituted in formula **1**. This gives

$$d = .136\sqrt{13 \times 21} = 2.247, \text{ say } 2\tfrac{1}{4}, \text{ in.}$$

Hence, the solution by formula **1** for low-speed stationary engines is satisfactory, and the value of d thus found may be considered as justified by practice and may be accepted. Ans.

53. Connection of Rod to Piston.—Modes of fastening the piston rod to the piston are shown in Figs. 13 to 18. The end of the rod is tapered and threaded to receive a nut or it is riveted. The taper may vary in different cases from $\frac{1}{2}$ to 3 inches per foot. The cross-section of the rod at root of threads should be such as to give a tensile strength of 5,000 pounds per square inch for wrought iron and 7,000 pounds for steel. Letting d_1 be the diameter of rod at root of thread,

$$.7854\, d_1^2 \times 5{,}000 = .7854\, D^2 p \text{ for wrought iron}$$

and

$$.7854\, d_1^2 \times 7{,}000 = .7854\, D^2 p \text{ for steel}$$

Or, $d_1 = .014\, D \sqrt{p}$ for wrought iron

and $d_1 = .012\, D \sqrt{p}$ for steel

Rods having collars forged on to bear against the piston are much to be preferred for heavy pressures. When the section of a rod is to be reduced, it is important, especially for a steel rod, to provide a liberal fillet, as shown in Fig. 14, so as to leave no sharp corners.

CONNECTING-ROD

CALCULATION FOR CONNECTING-ROD

54. The first dimension of the connecting-rod that should be determined is the length. A very short rod is objectionable, because the shorter the rod the greater is the thrust along it, due to the steam pressure on the piston, and the greater are the irregularities introduced into the steam distribution by the angularity of the connecting-rod. On the other hand, increasing the length of the connecting-rod increases the stresses in it, as a column in compression, so that an unnecessarily long rod should also be avoided.

As a working compromise, in stationary practice, the length of connecting-rod between centers of crankpin and wristpin is generally made between 5.5 and 6 times the length of the crank-arm; it is seldom that the connecting-rod of a stationary engine will go beyond these limits. However, if it is necessary to shorten the engine, the length of connecting-rod may be considerably less. In naval engines, the ratio of the connecting-rod to the crank is often as small as 4, and sometimes only 3.5.

The connecting-rod is subjected to alternate tension and compression; therefore, in design it is to be treated as a long column under compression. Considered as a column, the connecting-rod is pin-ended and guided at both ends in the plane of motion and is square-ended in the plane at right angles.

55. In low-speed engines, the connecting-rod frequently has a circular cross-section; and, in the older practice at least, was usually swelled from the necks near the stub ends to a maximum diameter at the middle. With a rod of this kind, the diameter at the middle, which will be the section of maximum stress, is given by the following formula, which is based on Euler's formula:

$$d = 2\sqrt[4]{\frac{f p D^2 L^2}{10 E}}, \qquad (1)$$

in which d = diameter of rod, in inches;

f = factor of safety, which for this kind of rod varies in good practice from about 8 to about 23, the average value being about 16;

p = maximum unbalanced steam pressure on piston, in pounds per square inch, and should not be less than 113;

D = diameter of cylinder, in inches;

L = length of connecting-rod from center of wrist-pin to center of crankpin, in inches;

E = modulus of elasticity of material, which for steel is 30,000,000.

With a given material and a definite steam pressure, the values of E, p, and f may be taken as constant, and formula **1** reduces to

$$d = K\sqrt{D L}, \qquad (2)$$

in which the value of K is $2\sqrt[4]{\frac{f p}{10 E}}$.

An examination of steel connecting-rods of this kind designed for low-speed stationary engines carrying a steam pressure of 100 pounds, gauge, shows that in good practice K has a value from .0816 to .105, with an average value of .0935.

Formula **1** may be used in the design of all connecting-rods of circular cross-section. Formula **2** may be used directly in the design of the class of rods for which the constant K is derived; but, for other rods, it should not be used except as a check on the results given by formula **1.**

56. The usual custom is to taper the rod, and if it is tapered both ways the diameter at the middle should be determined by formula **1,** Art. **55.** The diameters at the necks, next to the stub ends, are then usually calculated for simple tension and compression by the formula

$$d' = D\sqrt{\frac{p f}{S_c}},$$

in which d' = diameter at neck, in inches;

S_c = maximum compressive stress in material, in pounds per square inch, which for mild steel may be taken as 50,000.

The other symbols are the same as for formula **1,** Art. **55.** If, in any particular case, a value has been assigned to f, to be used in formula **1,** Art. **55,** the same value should be assigned to f in the formula given in this article. The rod is then given a uniform taper both ways, from middle to neck, and the corners are filleted or rounded off as required.

If the rod tapers from crosshead to crank end, which is preferable to having it taper both ways, except for low speeds, the diameter at the crosshead neck should be determined by the formula of this article. The diameter of the middle should be determined by formula **1,** Art. **55,** and the rod should then be given a uniform taper throughout.

57. In high-speed engines, the rod is often rectangular or approximately so in cross-section, and is usually deepest at the crank end. In this class of rods, the breadth at the middle, that is, the dimension perpendicular to the plane of motion of the rod, may be determined by the formula

$$b = .588\sqrt{\frac{f p}{E}}\sqrt{D L}, \qquad (1)$$

in which b = breadth of rod at middle, in inches;

f = factor of safety, which in good practice varies from about 9 to about 57, the average value being about 20.

The other symbols are the same as for formula **1,** Art. **55.**

Where f, p, and E may be taken as having definite values, formula **1** reduces to

$$b = K\sqrt{D\,L}, \qquad (2)$$

in which K represents $.588\sqrt[4]{\frac{f\,p}{E}}$.

An examination of steel connecting-rods of this class in a large number of high-speed stationary engines carrying steam pressures of 100 pounds, gauge, shows that K varies from .0433 to .0693, with an average value of .0545 in good practice.

In rods of this class, the breadth at the mid-section having been determined, the height at the mid-section, that is, the dimension parallel with the plane of motion, may be taken as from 2.18 to 4 times the breadth, the average ratio of height to breadth at mid-section being 2.75. The greater the piston speed and rotative speed, the greater should be the ratio of height to breadth at this section.

Formula **1** may be used in the design of all connecting-rods of rectangular cross-section. Formula **2** should be used directly only in the design of such rods as those from which the value of K is derived, and should not be used in other cases, except as a check on the results obtained by formula **1**.

58. While the formulas in Art. **57** give results agreeing quite closely with good practice, some investigators think it advisable to check the unit stress in a rod calculated by their use. This is done by means of the following formula, which is known as **Ritter's formula.** This check is suggested owing to the wide range of the factor of safety and the consequent danger of passing the limit of safety in the use of Euler's formula,

$$S = \frac{P}{A}\left(1 + \frac{S_e}{\pi^2\,m\,E} \times \frac{l^2}{r^2}\right),$$

in which S = greatest unit compressive stress in rod, in pounds;

P = maximum thrust on connecting-rod, in pounds;

A = area of cross-section of rod, in square inches;

S_e = unit stress of the material at elastic limit; usually taken as 26,000 for wrought iron and 35,000 for steel;

π^2 = 9.8696;

m = a constant = 4 for plane at right angles to plane of motion of rod;

E = modulus of elasticity, which is 25,000,000 for wrought iron and 30,000,000 for steel;

l = length of column, being taken in this case from center of crankpin to center of wristpin;

r = radius of gyration, being $\frac{1}{16}d^2$ for circular cross-sections of rod and $\frac{1}{12}a^2$ for rectangular cross-sections;

d = diameter of circular rod, in inches;

a = least dimension of rectangular rod, in inches.

The value of S found by the formula just given should not exceed the elastic limit as given for S_e.

59. The crosshead neck should be designed for simple tension and compression. The breadth should be retained as determined for the mid-section, and the height should be determined by the formula

$$h' = \frac{.7854\,D^2 p f}{b\,S_c},$$

in which h' = height of rod in crosshead end neck, in inches;

D = diameter of steam cylinder, in inches;

p = maximum unbalanced steam pressure on piston, in pounds per square inch, which should not be less than 113;

f = factor of safety, which, in any particular case, should be taken the same as is used in calculating the mid-section by formula **1**, Art. **57**;

b = breadth of rod in crosshead end neck, in inches;

S_c = maximum compressive stress in material, in pounds per square inch, which, for mild steel, may be taken as 50,000.

After the dimensions of these two cross-sections are determined, the rod is usually made of equal thickness throughout, and is given a uniform taper from the neck at the crosshead end to the crankpin end.

PROPORTIONS FOR CONNECTING-RODS

60. Solid and Open Connecting-Rod Ends.—The design of the ends of the rod is generally based on empirical expressions deduced from practice. The proportions that follow are based on the practice of the best engine builders. Fig. 22 shows a style of rod that gives excellent results in stationary work. The crosshead end is forged solid and cut out for the brasses, which are made without top and bottom flanges on one side, so that they can be slipped into the rod. The brasses are held in place and adjusted by the steel wedge w_1 and the adjusting screws S_1. The brasses on the crosshead ends of connecting-rods are seldom babbitted, as experience shows that it is unnecessary.

The crank end of the rod is made fork-shaped, so that the brasses can be put on the crankpin and then slipped into the rod from the end. If the wristpin cannot be removed from the crosshead, such a construction must also be used for the crosshead end of the rod. The bolt B, which is turned and fitted in a reamed hole, holds the brasses, which are adjusted by a steel wedge and screws, in the same manner as for the crosshead end. The crankpin brasses are babbitted, as shown.

61. The dimensions of this rod are to be made according to the following proportions, which are suitable for a rod of either steel or wrought iron.

For the wristpin, or crosshead pin, end:

D = diameter of cylinder, in inches.

d = diameter of crosshead pin, in inches, which will be treated in *Steam-Engine Design*, Part 2.

$a = 1.42\,d.$

$b = 1.125\,d + .375$ inch.

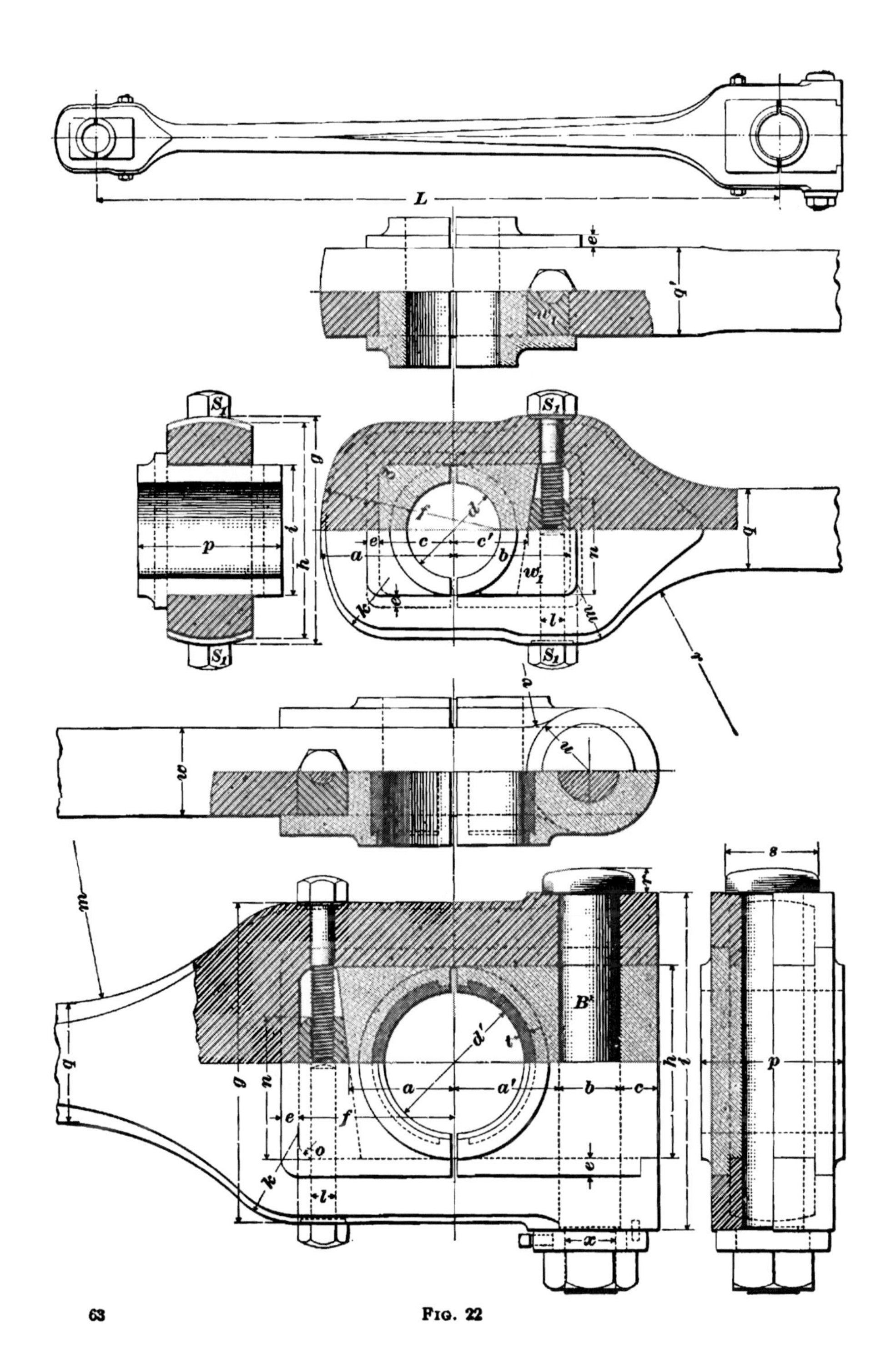

Fig. 22

$c = .75\,d + .125$ inch.
$c' = .75\,d + .125$ inch.
$e = .125\,d$.
$f = 1.92\,d$.
$g = 2.375\,d$.
$h = 2.25\,d$.
$i = 1.35\,d$.
$k = .625\,d$.
$l = .035\,D + .25$ inch.
$m = .625\,d$.
$n = d$.
$o = .125\,d$.

p = length of wristpin journal − .125 inch in small engines, and length of wristpin journal − .25 inch in large engines.
q should be designed by the formula of Art. **56**, q being the same as d' in that formula.
$q' = 1.1\,q$.
$r = 1.75\,d$.

Taper of adjusting wedges = $1\frac{1}{2}$ inches per foot.

For the crankpin end:

D = diameter of cylinder, in inches.
d' should be designed by the rules for finding the diameter of crankpin, according to Arts. **34** to **37**.
$a = .75\,d'$.
$a' = .75\,d'$.
$b = .1\,D + .4375$ inch.
$c = .625\,b$.
$e = .125\,d'$.
$f = d' + .5$ inch.
$g = 2.25\,d'$.
$h = 1.35\,d'$.
$i = 2.375\,d'$.
$k = .625\,d'$.
$l = .035\,D + .25$ inch.
$m = 2\,d'$.
$n = d'$.
$o = .125\,d'$.

p = length of crankpin journal − .125 inch in small engines, and length of crankpin journal − .25 inch in large engines.
q should be determined by the formulas given for designing the crankpin neck of the connecting-rod in Art. **56**.
$r = .375\,b$.
$s = 1.5\,b$.
$t = .02\,D + .0625$ inch.
$u = b$.
$v = b$.
w should be designed by formula **1**, Art. **57**, w being the same as b of that formula.
$x = .8\,b$.

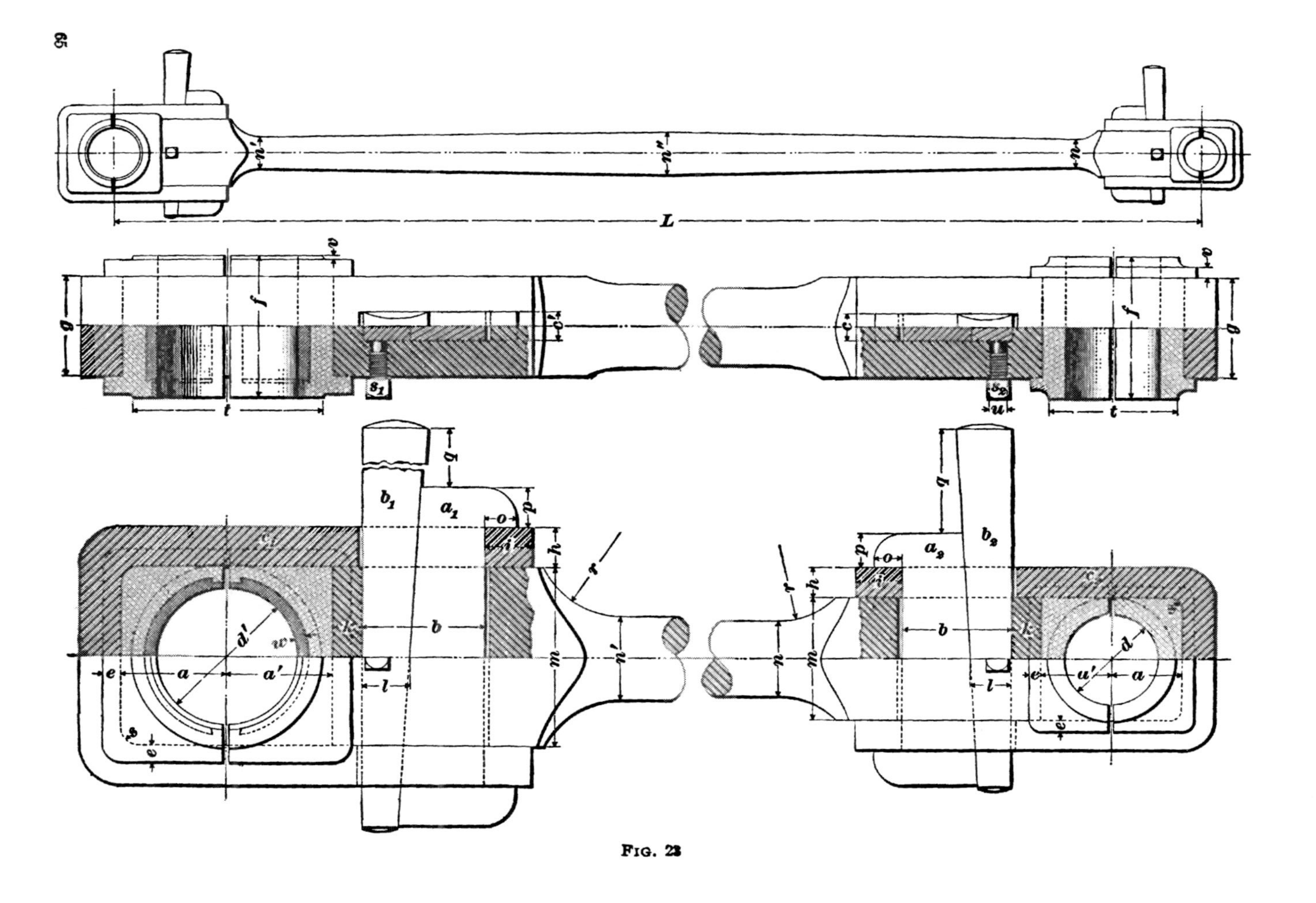

FIG. 23

In designing the cross-section of the rod at the middle of its length, the breadth should be determined by formula **1**, Art. **57**. At the point where the breadth is measured, the corners of the cross-section should not be rounded off enough to reduce the height to less than twice the breadth. The total height should be from 2.18 to 4 times the breadth, the average being 2.75.

The nut and locking collar should be designed according to the principles of machine design.

The taper of the adjusting wedge may be made $1\frac{1}{2}$ inches per foot.

62. Strap-End Connecting-Rod.—In Fig. 23 is shown a **strap-end connecting-rod.** This rod is commonly used and has been found very satisfactory. However, the strap end should not be used on the crankpin, where either the piston speed or the rotative speed is high. The straps c_1 and c_2 are fastened to the ends of the rods by means of gibs a_1 and a_2 and cotters b_1 and b_2. The cotters are held in place by setscrews s_1 and s_2. Small steel blocks, shown between the ends of the setscrews and the cotters, are used to prevent the cotter from being injured by the setscrews.

The rod, cotters, gibs, and straps are generally made of steel, but wrought iron is occasionally employed. The crankpin brasses are shown babbitted, and the wristpin brasses without Babbitt. The brasses are adjusted by means of the cotters, which draw the straps farther on to the rod when they are driven in.

The dimensions for the rod shown in Fig. 23 are given by the following proportions:

For the wristpin end:

D = diameter of cylinder, in inches.

d = diameter of wristpin, in inches, which will be treated in *Steam-Engine Design*, Part 2.

$x = .7854\, n^2$ = a factor for use in finding proportions that follow.

$a = .75\, d + .125$ inch.

$a' = .75\, d + .125$ inch.

$b = \sqrt{2.5\, x}$.

$c = .25\,b.$

$e = .125\,d.$

$f =$ length of wristpin journal − .125 inch in small engines, and length of wristpin journal − .25 inch in large engines.

$g = 1.3\,n.$

$h = \dfrac{.5\,x}{g - c}.$

$i = \dfrac{.32\,x}{h}.$

$k = \dfrac{x}{1.8\,d}.$

$l = .375\,b.$

$m = 1.35\,d$ for wristpins up to 3.5 inches in diameter.

$m = 1.48\,n$ for pins above 3.5 inches in diameter.

n should be determined by the formula of Art. **56,** being the same as d' of that formula.

$o = .25\,b.$

$p = .33\,b.$

$q = 1.125\,d$ for wristpins up to 3.5 inches in diameter, and $q = 4$ inches, constant, for pins above 3.5 inches in diameter.

$r = n.$

$s = .125\,d.$

$t = 1.35\,d.$

$u = .02\,D + .25$ inch.

$v = .125\,d.$

The taper of the cotter is $\frac{3}{4}$ inch per foot.

For the crankpin end:

$D =$ diameter of cylinder, in inches.

d' should be determined by the rules for finding the diameter of crankpins, given in Arts. **34** to **37.**

n' should be found by the formula of Art. **56,** being the same as d' of that formula. In practice, the value of d' found by the formula is often multiplied by 1.1, to find the value of n', Fig. 23. Though, theoretically, this increase in the diameter of the neck is unnecessary, it may be considered good practice as an empirical allowance against unknown stresses due to inertia and seizing of the crankpin brasses from insufficient lubrication.

$x' = .7854\,n'^2 =$ a factor used in the proportions that follow.

$a = .75\,d'.$

$a' = .75\,d'.$

$b = \sqrt{2.5\,x'}.$

$c' = .25\,b.$

$e = .125\,d'$.

f = length of crankpin journal − .125 inch, in small engines, and length of crankpin journal − .25 inch in large engines.

$g = 1.3\,n$, the same as for wristpin end.

$h = \frac{.5\,x'}{g - c'}$.

$i = \frac{.32\,x'}{h}$.

$k = \frac{x}{1.8\,d'}$, the same as for wristpin end.

$l = .375\,b$.

$m = 1.3\,d'$.

$o = .25\,b$.

$p = .33\,b$.

q = same values as for wristpin end.

$r = 1.1\,n$.

$s = .25\,d'$.

$t = 1.35\,d'$.

v = .125 inch (constant).

$w = .02\,D + .0625$ inch.

n'' should be found by formula **1**, Art. **55**, being the same as d of that formula.

Taper of cotter = $\frac{3}{4}$ inch per foot.

63. Marine-End Connecting-Rod.—Fig. 24 shows a connecting-rod with **marine ends.** For marine engines, and in some cases for stationary engines, the crosshead end is forked as shown. For most stationary work, however, the crosshead end is not forked, and a solid or strap end is often used with a marine crankpin end. The object in using the forked end is to obtain the most solid and substantial form of crosshead. Where this is not necessary, the forked-end rod should be avoided, because it is impossible to adjust the wristpin brasses so accurately as to divide the pressure equally between them. The inequality of pressure on the wristpins brings a side thrust on the crosshead and thus increases the stresses in the connecting-rod.

In the connecting-rod shown in Fig. 24 (a), the brasses are held to the ends of the rod, which are forged to a **T** shape, by turned bolts. These bolts pass through steel or wrought-iron caps and the brasses. Liners are fitted between the two parts of each set of brasses, and when the brasses become worn, the liners are taken out and either filed or planed down, thus allowing the brasses to be tightened.

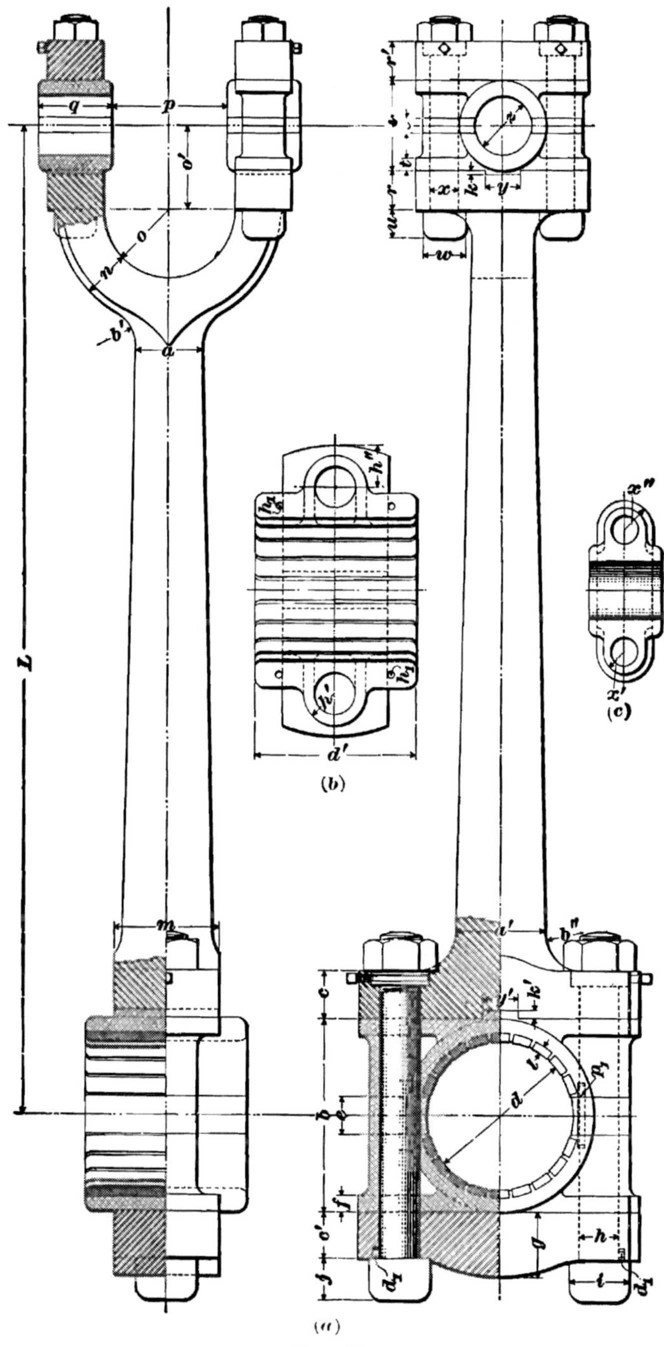

FIG. 24

The crankpin brasses are babbitted, the Babbitt being poured so as to project about $\frac{1}{8}$ inch above the brass in order that the pin will bear only on the Babbitt, with oilways between the Babbitt blocks.

The liners in the crankpin end are sometimes secured by small pins p_1, which pass through the liner and project into the holes h_1, shown in the view of the brass given in Fig. 24 (*b*). Fig. 24 (*c*) shows a view of one of the wristpin brasses. The rod and caps are usually made of steel, although wrought iron may be used.

64. The dimensions for the rod shown in Fig. 24 (*a*) are determined by the following proportions:

a = diameter of rod at wristpin end, and should be calculated by the formula of Art. **56**, a being the same as d' of that formula. In marine work, good material and workmanship should be employed, so that a small factor of safety can be used in design, as it is imperative to reduce weight.

a' = diameter of rod at crankpin end, and should be determined according to Art. **56**.

$b = 1.3\,d$.

$b' = .6\,a$.

$b'' = .5\,a'$.

$c = .7\,a$.

$c' = .7\,a$.

d = diameter of crankpin.

d' = length of crankpin journal − .125 inch.

$e = .25\,d$.

$f = .35\,c$.

$g = 1.4\,c$.

h = diameter of bolt at root of thread, when area of cross-section at root of thread is $.00008\,A\,P$ for steel and $.0001\,A\,P$ for wrought iron, where A is the area of cylinder, in square inches, and P the boiler pressure, in pounds per square inch, which should not be less than **113**.

$h' = .875\,h$.

$h'' = h$.

$i = 1.5\,h.$

$j = h.$

$k = .05\,z + .0625$ inch.

$k' = l.$

$l = .05\,d + .0625$ inch.

$m = 1.1\,a'.$

$n = .7\,a.$

$o = .5\,p + .5\,(q - 2\,x).$

$o' = .6\,a + .675\,z.$

$p = 2\,B + .25$ inch, where B is the diameter of piston rod.

$q =$ length of wristpin as made for crosshead.

$r = .6\,a.$

$r' = .6\,a.$

$s = 1.68\,z.$

$t = .35\,r.$

$u = x.$

$v = .25\,z.$

$w = 1.5\,x.$

$x =$ diameter of bolt at root of thread, when area of cross-section is $.00004\,AP$ for steel and $.00005\,AP$ for wrought iron.

$y = .25\,d.$

$y' = .25\,d.$

$z =$ diameter of wristpin as made for a solid crosshead.

$x' = .875\,x.$

$x'' = x.$

Proportions for locknuts should be determined according to the principles of machine design. The dowel-pins shown at d, Fig. 24 (a), are used to keep the bolt from turning with the nut. The distance between the bolts should be such that the brass will be about .25 inch thick at the thinnest part between the bolt and the pin.

65. In marine design, all the connecting-rods for the main engines should be made alike, so that they may be interchangeable. This is done so that a large quantity of duplicate parts will not have to be carried. In the case of a marine engine with several cylinders, the designs should be worked out for all the cylinders, and the heaviest design found necessary for any cylinder should be adopted for all. The same rule applies to all marine-engine parts that can be made interchangeable.

Merchant Books

STEAM-ENGINE DESIGN

(PART 2)

ENGINE DETAILS

CROSSHEADS

1. Corliss Engine Crosshead.—Crossheads are made in a great variety of forms, some of the most important of which, together with the proportions for their design, are given in the following pages. Fig. 1 shows a crosshead much used on Corliss and similar engines. This crosshead is composed of a cast-iron box, with a boss for a piston rod and bosses for the wristpin. The wristpin is generally made of steel—occasionally of wrought iron—with the ends tapered so as to fit snugly into the crosshead. This pin is held by a nut and washer and has a projecting part y, which is drilled for an oil cup. Holes are drilled into the pin, as shown by the dotted lines, for the purpose of leading oil to the connecting-rod brasses.

2. Shoes or gibs g_1 are fitted to the crosshead on tapered ways, and these gibs can be adjusted by means of the bolts t_1. The gibs shown on the crosshead have cylindrical bearing surfaces, but the cross-section of a gib with the bearing surface **V**-shaped is shown at A. The engine guides are made either cylindrical or **V**-shaped, to correspond with the gibs.

Let s = diameter of wristpin, in inches;

t = length of wristpin journal, in inches.

Then, s and t are given by the following formulas:

$$s = \sqrt[3]{\frac{t D^2 p}{S_4}}, \qquad (1)$$

$$s t = \frac{.7854 D^2 p}{P}; \qquad (2)$$

in which D = diameter of cylinder, in inches;

p = maximum unbalanced pressure on the piston, in pounds per square inch;

S_4 = safe bending stress in the material of the pin, in pounds per square inch; as the crosshead pin, or wristpin, works under a rapidly reversing load and is also subject to considerable shock, S_4 should not exceed about 7,000, except in case of unusually good material and workmanship and careful inspection, when it may equal 8,000;

P = maximum allowable pressure on the bearing, in pounds per square inch of projected area. P may have a value from 800 to 1,400, but the best practice gives 1,200 to 1,350 as the highest values, the lower values of P corresponding to the higher rotative speeds.

If p and P have definite values, formula **2** may be written

$$s t = K D^2, \qquad (3)$$

in which K is a constant representing $\frac{.7854 p}{P}$.

An examination of the wristpins of a large number of low-speed, side-crank stationary engines, in which the type of crosshead shown in Fig. 1 is generally used, shows K to have a value ranging from .042 to .083, with an average value of .058 in good practice. The value of t in terms of s may be found by formula **2.** By substituting this value of t in formula **1,** s may be found, after which t may be definitely determined by substituting the value of s in formula **2.** Formulas **1** and **2** apply to all wristpins supported at both ends.

3. Let f represent the width of the shoe, in inches, as shown in Fig. 1, and g the length of the shoe, in inches.

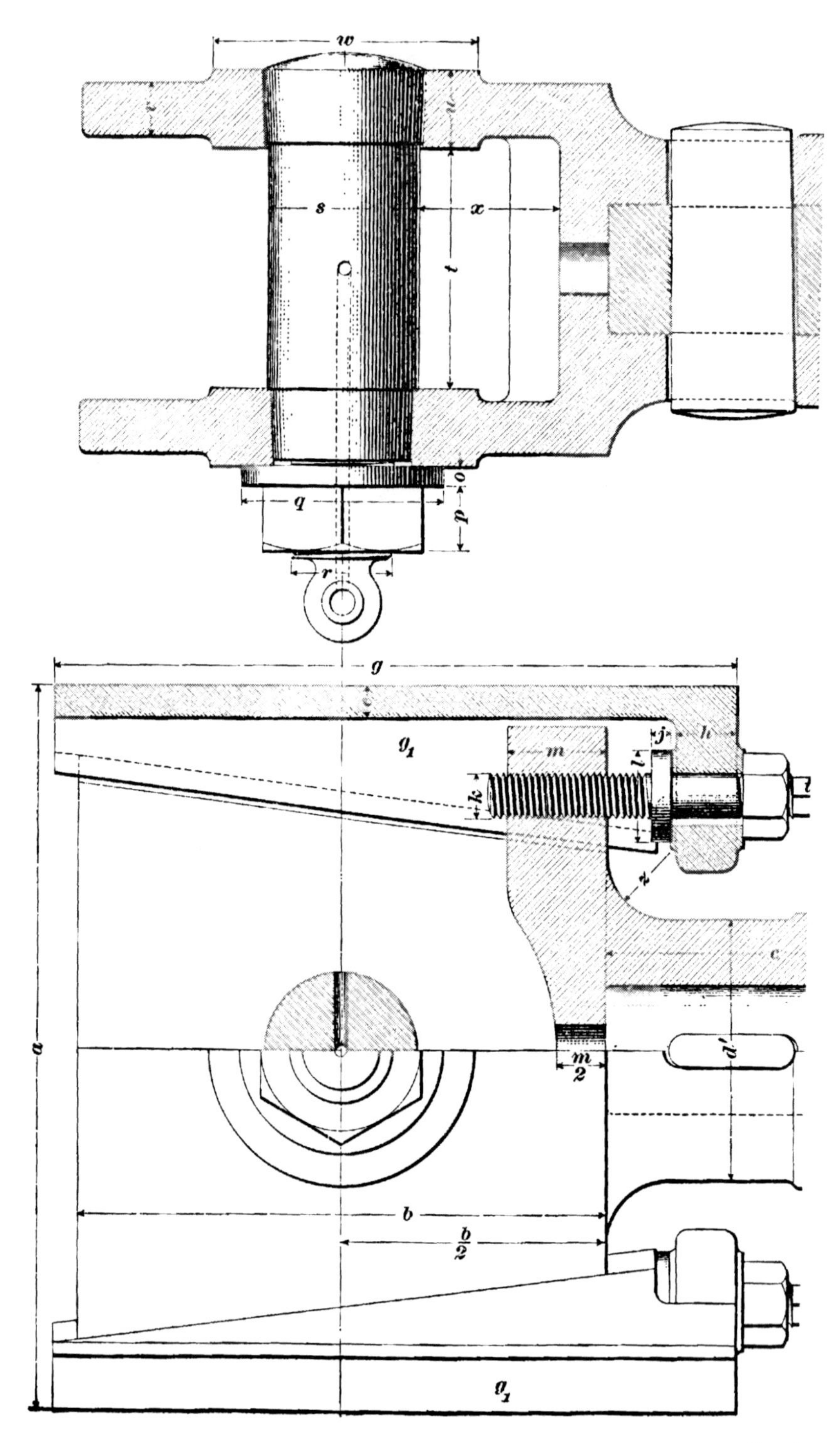

FIG 1

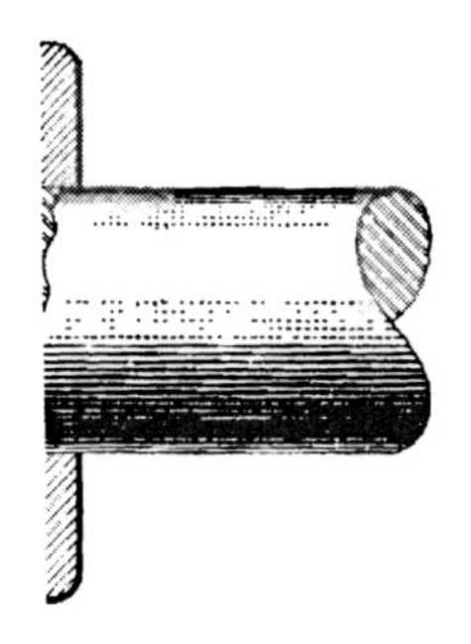

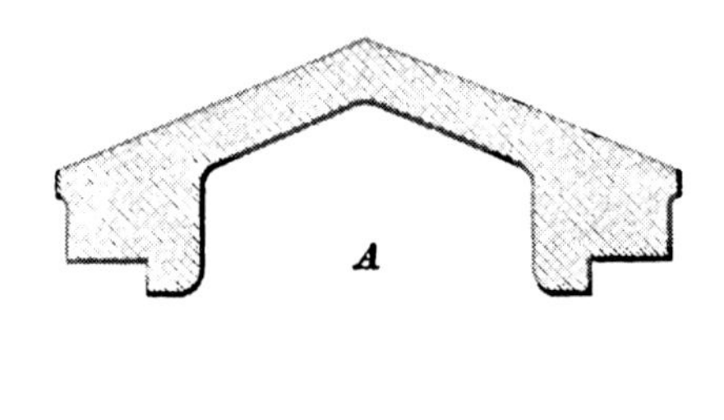
A

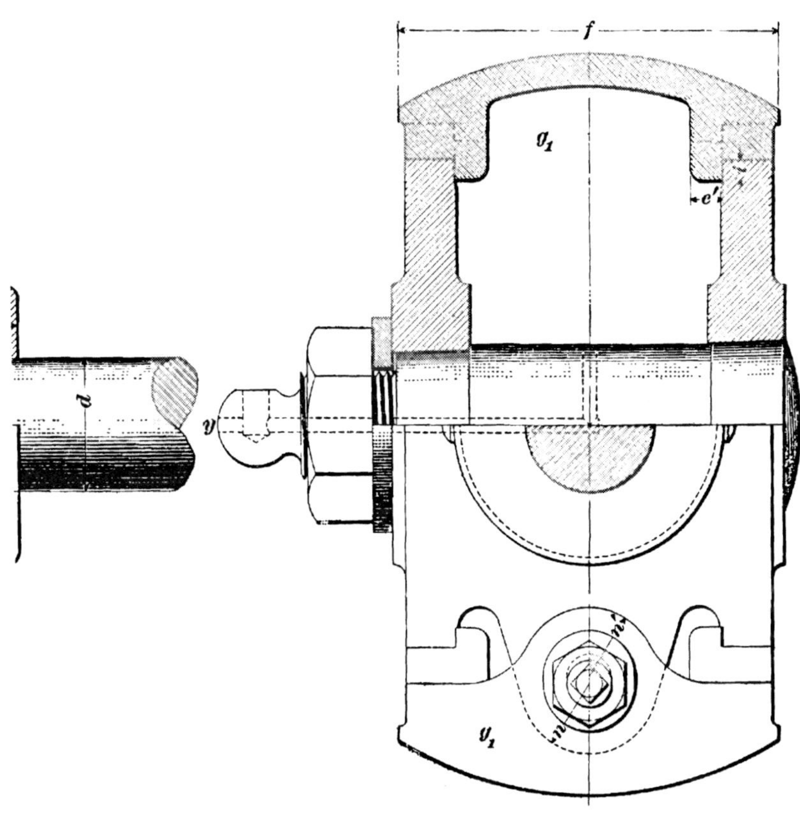
f
g_1
e'
d
y
n
m
g_1

Then, for cylinders up to and including 26 inches in diameter,

$$f = .357\,D + 1.625 \text{ inches}$$

For cylinders larger than 26 inches in diameter,

$$f = .376\,D + 1 \text{ inch}$$

The length of the shoe g is given by the formula

$$g = .7854\,\frac{D^2 p}{f P \sqrt{n^2 - 1}}, \qquad (1)$$

in which P = maximum allowable pressure on the guide, in pounds per square inch;

n = ratio of length of connecting-rod between center of crankpin and center of wristpin to length of crank between center of crankpin and center of shaft;

D = diameter of cylinder, in inches;

p = maximum unbalanced steam pressure, in pounds per square inch, which should not be taken less than 113.

An examination of a large number of engines shows the average value of P to be 36 for low-speed stationary engines and 27 for high-speed engines. Some authorities give values varying from 40 to 100, but the values just stated represent the best American practice, and as these values are safer, they are to be preferred.

In low-speed stationary engines, the value of n is about constant. Then, in this class of engines, with definite values for p and P, formula **1** reduces to

$$f g = K D^2 \qquad (2)$$

An examination of low-speed stationary engines shows the value of K to vary from .23 to .52, with a value of .37 as the average in good practice. The higher the piston speed, the larger the value that should be assigned to fg.

4. The remaining dimensions of the crosshead shown in Fig. 1, are given by the following proportions:

D = diameter of cylinder.

d = diameter of piston rod.

a must be found by making a scale drawing of the connecting-rod in its extreme position. Care must be taken that the rod will clear the gibs, and also the guides, for all positions.

b = length of crosshead body, and must be given such a length that $\frac{b}{2}$ will accommodate the end of the connecting-rod to be used. This length can best be determined on a drawing board.

$c = 2.125\,d$.

$d' = 2\,d$.

$d'' = 2.125\,d$.

$e = .01\,D + .5$ inch.

$e' = e$.

$h = .04\,D + .5625$ inch.

$i = .6\,e$.

$j = .015\,D + .125$ inch.

$k = .04\,D + .3125$ inch.

$l = .08\,D + .625$ inch.

$m = .12\,D + .125$ inch.

$n = .06\,D + .5$ inch.

$n' = n$.

$o = .19\,r$.

$p = .625\,r$.

$q = 2.125\,r$.

$r = .66\,s$.

s = diameter of wristpin, in inches (see Art. 2).

t = length of wristpin journal, in inches (see Art. 2).

$u = .05\,D + .625$ inch.

$v = .043\,D + .3125$ inch.

$w = 1.75\,s$.

x must be determined by making a scale drawing of the wristpin end of the connecting-rod, and must be such that the connecting-rod will swing clear of the crosshead in all positions by at least $\frac{1}{8}$ inch.

y is plug for oil hole.

$z = .5\,D$.

Taper of gibs, 1.5 inches per foot.

For cylinders above 20 inches in diameter, the gibs should be ribbed.

This crosshead is designed to be used with a solid-end connecting-rod.

5. Marine-Engine Crosshead.—Fig. 2 shows a style of crosshead used mostly for marine work. The two wristpins a_1, a_1 and the block b_1 are one solid steel forging. This crosshead requires a forked-end connecting-rod. The bearing surfaces are composed of two cast-iron shoes fastened to the block by bolts. These shoes are babbitted, the Babbitt

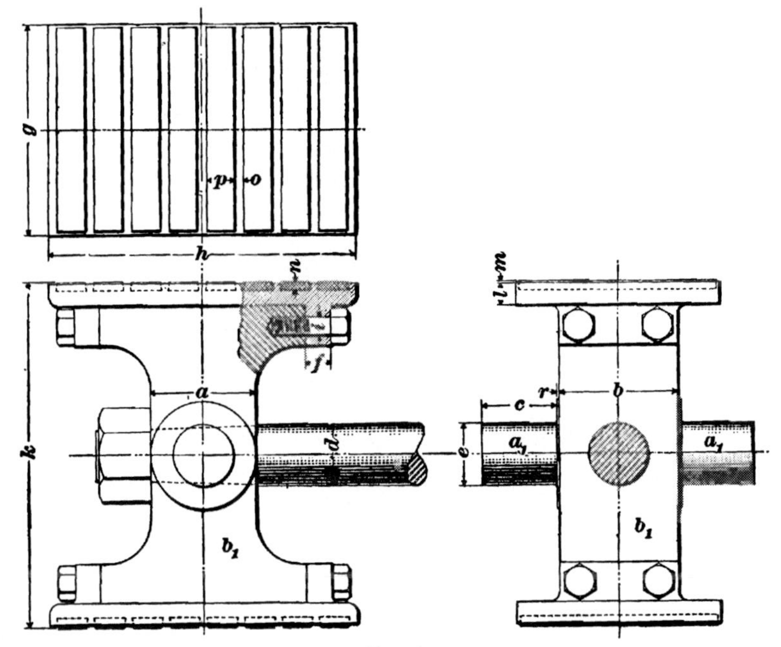

FIG. 2

being dovetailed into and raised a little above the surface of the iron, so that no wear will come directly on the iron. The piston rod is tapered in the block and fastened with a nut.

Let e = diameter of wristpins, as shown in Fig. 2;
c = length of each wristpin.

Then, e and c are given by the following formulas:

$$e = \sqrt[3]{\frac{2.67\ D^2 p c}{S_4}}, \qquad (1)$$

$$ec = \frac{.524\ D^2 p}{P}; \qquad (2)$$

in which D = diameter of cylinder, in inches;

p = maximum unbalanced steam pressure on the piston, in pounds per square inch, which should not be taken less than 113;

S_4 = safe bending stress in the material of the wrist-pin, in pounds per square inch, which, for marine work may be taken as high as 12,500; if the crosshead should be designed for a stationary engine, S_4 had better be kept at a value not greater than 7,000;

P = maximum allowable pressure of the wristpin brass on the wristpin, in pounds per square inch of projected area. For a marine design, P may be made 1,350; for a stationary design, P had better be kept down to 1,200.

6. Let g represent the width of the crosshead shoe, as shown in Fig. 2, and h the length of the crosshead shoe. Then, g is given by the formula

$$g = .11\, D\sqrt{\frac{p}{n}}, \qquad (1)$$

in which n is the ratio of length of connecting-rod to length of crank. The other symbols are the same as in Art. **5.**

The length h is obtained by the formula

$$h = \frac{.7854\, D^2 p}{g P \sqrt{n^2 - 1}}, \qquad (2)$$

in which P is the maximum allowable pressure of the cross-head shoe on the guide, in pounds per square inch. For a marine design, P may be taken at from 40 to 100 on the guide on which the pressure comes when the vessel is going ahead, and as high as 400 on the guide on which the pressure comes when the vessel is backing, if it is important to save weight. For a stationary engine, P had better be kept at a much lower value, not to exceed 30, and as low as 16, if possible. In all cases, the higher the piston speed, the lower should be the value of P.

7. The remaining dimensions of this crosshead are based on the following proportions:

D = diameter of cylinder.
d = diameter of piston rod.
$a = 1.75\,d$.
$b = 2\,d$.
$f = .05\,D + .5$ inch.
$i = \frac{3}{4}$ inch for $D = 15$ inches or less; $\frac{7}{8}$ inch for D = from 15 to 20 inches; 1 inch for D = from 20 to 25 inches, and $1\frac{1}{4}$ inches for D above 25 inches.
k must be such that the connecting-rod will clear crosshead and guides.
$l = .05\,D + .25$ inch.
$m = .125$ inch, constant.
$n = .016\,D$.
$o = .5$ inch, about.
$p = 1.75$ inches, about.
$r = .125$ inch, constant.

8. Fig. 3 shows a modification of the marine crosshead,

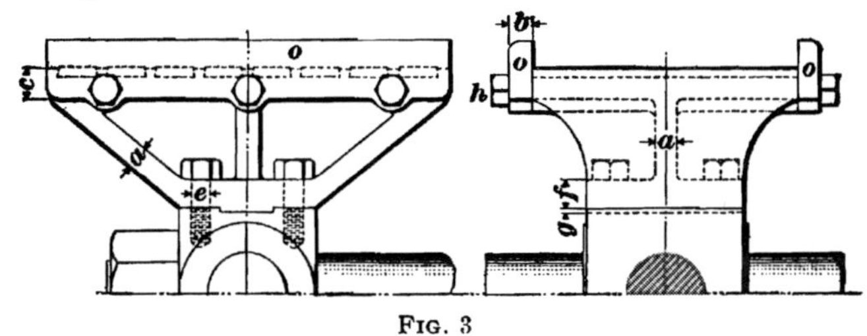

Fig. 3

in which the cast-iron shoe forms a much larger proportion and is provided with guiding strips o, o.

The proportions that apply to this form are:
$a = .03\,D + .5$ inch.
$b = .03\,D + .5$ inch.
$c = .05\,D + .5$ inch.
$e = \frac{7}{8}$ inch up to $D = 20$ inches; 1 inch up to $D = 25$ inches; and $1\frac{1}{4}$ inch for D above 25 inches.
$f = .05\,D + .5$ inch.
$g = .02\,D$.
$h = \frac{3}{4}$ inch for $D = 20$ inches, or less; $\frac{7}{8}$ inch for $D = 20$ to 25 inches, and 1 inch for D above 25 inches in diameter. Space bolts h not over 7 inches apart.

The other dimensions of this style of crosshead may be obtained from the proportions given for Fig. 2.

The shoes for the crossheads, Figs. 2 and 3, are adjusted by placing liners behind them.

9. Box-Bed Engine Crosshead.—Fig. 4 is an example of a crosshead that is much used on *box-bed engines*. The main part *A* is made of cast iron and has a boss into which the end of the piston rod is secured by means of a cotter.

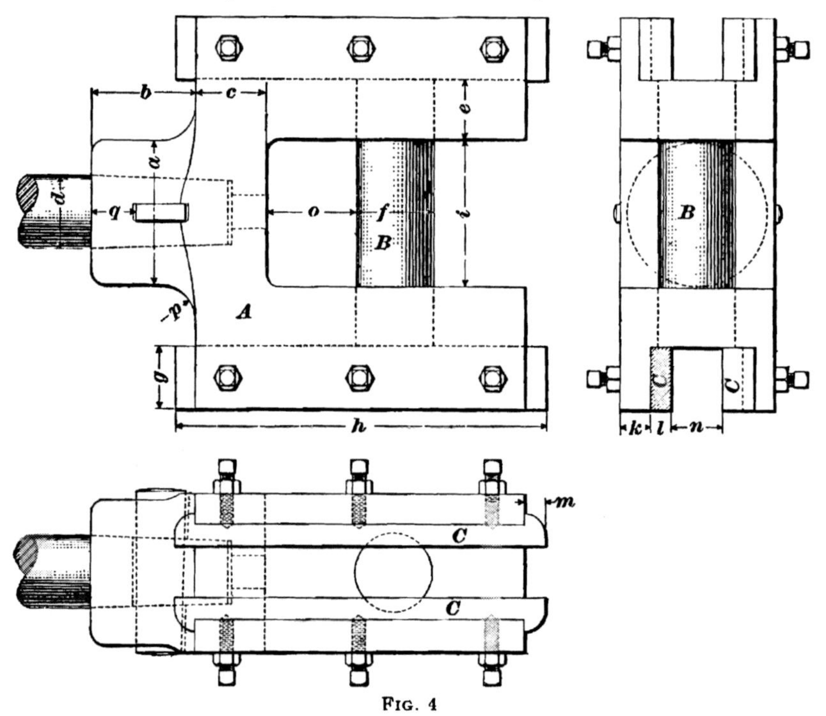

Fig. 4

The wristpin *B*, which is forced into the crosshead, is usually made of steel, but sometimes of wrought iron. Brass or bronze gibs *C*, which may be adjusted by means of the set-screws, furnish the surfaces that bear on the guides. The cotters should be designed according to the principles of machine design.

Let *f* represent the diameter of the wristpin, in inches, as shown in Fig. 4, and *i* the length of the wristpin journal, in inches. Then, for the diameter and length of a wristpin journal, f and i should be designed according to formulas **1** and **2**, Art. **2**.

The box-bed engine crosshead is used with a large range of speeds. When used with a low-speed engine, all the values may be taken exactly as given by formulas **1** and **2**, Art. **2**, but when used with a high-speed engine, the average value of P in good practice is only about 1,075. An examination of a large number of high-speed stationary engines shows that formula **3**, Art. **2**, applies to this case when the value of K varies between .0518 and .272, with an average value of .0825 in good practice.

10. Let g represent the width of the gib C, in inches, as shown in Fig. 4, which may be taken as $\frac{h}{4}$, and h the length of the gib C, in inches. Then,

$$h^2 = \frac{1.57\,D^2 p}{P\sqrt{n^2 - 1}}$$

The symbols are the same as given in Arts. **3** and **6**.

The following additional values of P are given for use with the formula just stated: With low-speed stationary engines, P varies from about 26 to 58 in good practice, the average being about 36. With high-speed stationary engines, P should be kept down to about 16, if possible. This will often involve making the crosshead excessively large, and in order to avoid this P may be carried up to 25 or even 30, if necessary. With locomotive engines, P varies from 40 to 60. In all cases, the higher the piston speed, the lower should be the value of P.

The other proportions for this type of crosshead are:

D = diameter of cylinder.
d = diameter of piston rod.
$a = 2\,d$.
$b = 1.5\,d$.
$c = d$.
$e = .75\,f$.
$k = .075\,D$.
$l = \frac{h}{18}$.
$m = l$.
n = thickness of guides.
o = space to clear connecting-rod.
$p = .5\,d$.
$q = .75\,d$.

Setscrews may be $\frac{3}{8}$ inch for cylinders up to 8 inches in diameter; $\frac{7}{16}$ inch for cylinders from 8 to 12 inches in diameter; and $\frac{1}{2}$ inch for all sizes above. Two setscrews should be used for cylinders up to 8 inches in diameter, and three for larger cylinders.

VALVES, VALVE STEMS, AND ECCENTRIC RODS

11. Valve Seats.—The slide valve has been fully considered in *Valve Gears*. Hence, it is necessary here to give only the proportions of the various parts. Two sectional views of a slide valve are shown in Fig. 5. The design of the valve seat will be considered first. The product of the width b and the length l of a steam port is obtained from the formula

$$l\,b = \frac{A\,S}{v},$$

in which A = area of piston, in square inches;
S = average piston speed, in feet per minute;
v = average velocity of steam, in feet per minute.

The width of the bridges c, c between the ports is usually made equal to the thickness of the cylinder walls. The width of the exhaust port is from $1\frac{3}{8}$ to $2\frac{1}{2}$ times the width of the steam port. In any given case, the exhaust port must be wide enough so that when the valve is at the end of its travel the width of the portion of exhaust port remaining open is at least equal to the width of the steam port.

12. Let a = width of exhaust port;
b = width of steam port;
c = width of bridge;
k = half of travel of valve;
i = inside lap;
o = outside lap.

Then, in order that the foregoing condition may be fulfilled, a should be equal to or greater than $b + k + i - c$. In Fig. 5,

$$h = a + 2\,(c - i)$$
$$e = b + i + o$$
$$L = h + 2e = a + 2c + 2b + 2o$$
$$B = l + 2t$$

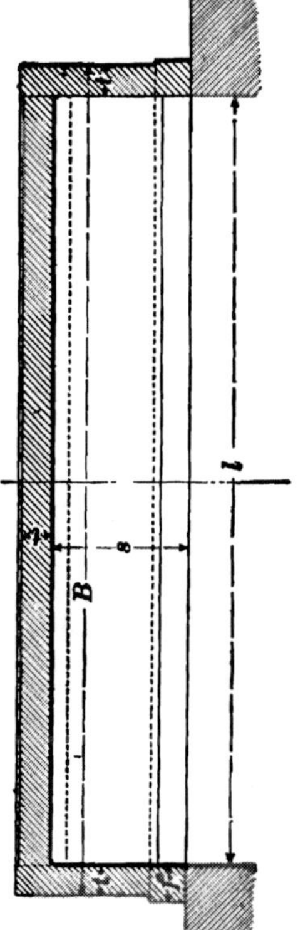

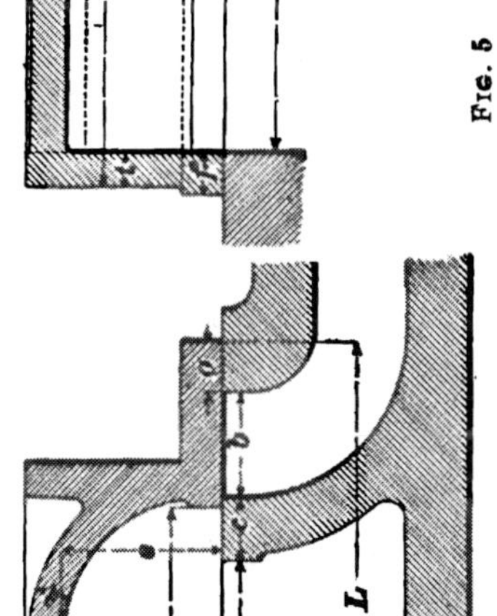

Fig. 5

When the valve at the end of its travel just opens the steam port fully, $k = b + o$.

The height s of the hollow underneath the valve must be sufficient to allow a free exhaust. For low piston speeds, s may be made equal to or slightly greater than the width b of the steam port; that is, $s = b$. More often, $s = \frac{3}{4} a$ to a, where a is the width of the exhaust port. Also, d may be $1.2\,t$, and f may be $1.1\,t$.

The thickness t of the metal of the valve equals

$t = .03\,D + .25$ inch (cast iron),

where D is the cylinder diameter, in inches.

The lead, lap, valve travel, etc. are readily determined from the valve diagram (see *Valve Gears*)

These quantities, in addition to the proportions just given, furnish sufficient data to design a plain slide valve, and may also be readily applied to the design of more complicated forms of slide valves, such as the double-ported valve, piston valve, etc.

13. The Valve Stem.—The valve stem must be designed to move the valve under the most unfavorable conditions that may occur in practice; hence, it may be assumed that the valve is unbalanced, for, even if balanced, the joint may leak. Furthermore, the valve may run dry on the seat, thus increasing the friction.

Let B = breadth of valve, in inches, as shown in Fig. 5;
d = diameter of valve stem, in inches;
E = modulus of elasticity of valve-stem material;
f' = coefficient of friction;
L = length of valve, in inches;
l = length of valve stem, in inches;
p = pressure on back of valve, in pounds per square inch, which may be taken as absolute boiler pressure;
f = factor of safety.

Then the load that the valve stem must move is equal to $f' p B L$ pounds.

Under the most favorable circumstances, f' should not exceed .2, but for designing purposes, f' should be made .25, as the lubrication may fail. Then, the load on the valve stem is .25 $p B L$ pounds.

As the valve stem is alternately under tension and compression, it should be designed as a column under compression and should be considered as pin-ended at both ends.

The diameter of the valve stem is therefore given by the formula

$$d = .85 \sqrt[4]{\frac{p B L l^2 f}{E}} \qquad (1)$$

As the load on the valve stem is generally much less than that just assumed, except under extraordinary conditions, f may be given a value of about 6. Then, taking E as 30,000,000 for steel and 25,000,000 for wrought iron, formula **1** becomes

$$d = K \sqrt[4]{p B L l^2}, \qquad (2)$$

in which K is equal to .018 for steel and .0194 for wrought iron.

EXAMPLE.—Find the diameter of a steel valve stem 17 inches long for a locomotive valve 18.5 in. × 10.25 in. on the face, the boiler pressure being 150 pounds, gauge.

SOLUTION.—Applying formula **2**,

$$d = .018 \sqrt[4]{164.7 \times 18.5 \times 10.25 \times 17^2} = 1 \text{ in., nearly. Ans.}$$

14. Valve-Stem Fastening.—Valve stems are fastened to valves in different ways. Fig. 6 shows a steel yoke suitable for a simple slide valve like that shown in Fig. 5.

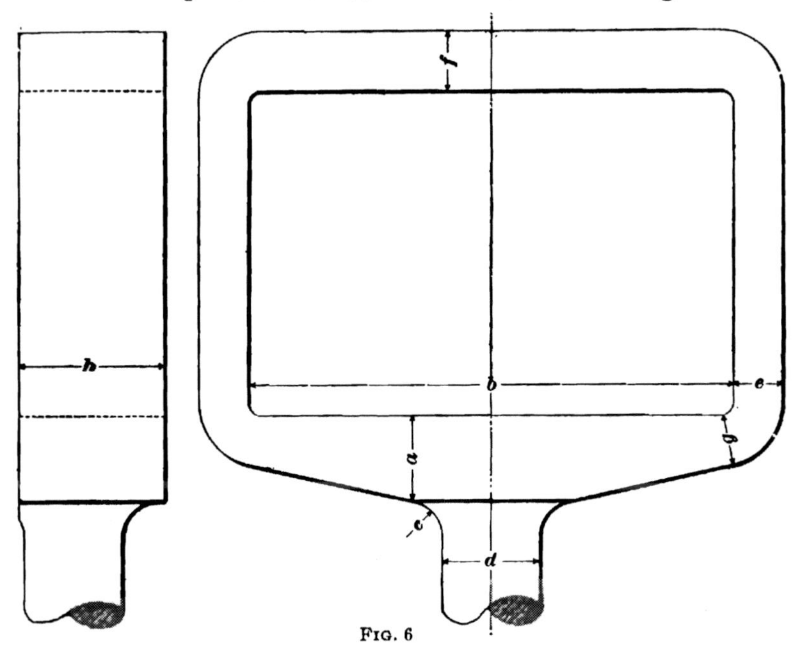

Fig. 6

The following proportions will show the proper dimensions. Where the meaning of a symbol is not stated, it is the same as given in Art. **13.**

$a = \frac{4f}{3}$.

b = dimension to fit valve.

$c = .375\,d$.

d = diameter of valve stem, in inches.

$e = \frac{pBL}{24{,}000\,h}$.

$f = .0025\sqrt{\frac{pBbL}{h}}$.

$g = .5\,d$.

h = dimension to fit valve.

15. Two other methods of fastening the valve to the stem are shown in Fig. 7. In (a), the valve stem is forged with two collars, and in (b) the end of the stem is threaded for two sets of nuts. Either arrangement allows the valve to adjust itself somewhat and thus prevents the stuffingbox from

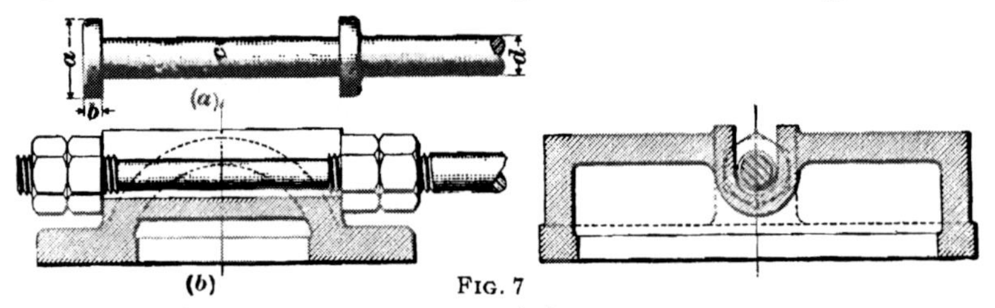

FIG. 7

wearing. The arrangement in (b) also furnishes a means for setting the valve, so as to get the proper location over the ports.

The proportions for the valve-stem fastenings shown in Fig. 7 are:

$c = d.$

$a = 2d.$

$$b = \frac{pBL}{62{,}800\,c}.$$

$d =$ diameter of valve stem.

The values of p, B, and L are as stated in Art. **13.**

In the arrangement shown in Fig. 7 (b), the thickness of each nut should be at least as great as b. The outer end of the valve stem may terminate in some form of crosshead running in guides or it may be jointed to a rocker-arm.

16. Eccentric Rods.—Eccentric rods may be rectangular or circular in cross-section. Quite often the rod is tapered, being largest where it joins the eccentric strap. It is common practice to make the area of the smallest section of the rod equal to about .8 the area of cross-section of the valve stem, the latter being calculated by either formula **1** or formula **2,** Art. **13.** The area at the large end may then be made about one-third larger than the area at the small end.

Fig. 8 shows an eccentric rod with right and left threaded ends, one of which is attached to the eccentric strap, and the other to a brass bearing for connecting to the valve-rod pin or rocker-arm pin. The threaded ends of the rod furnish means for adjusting the valve, and the locknuts N prevent the rod from turning after the valve is properly set.

The brass has a loose piece O, which can be adjusted by the cotter, thus furnishing means of taking up wear.

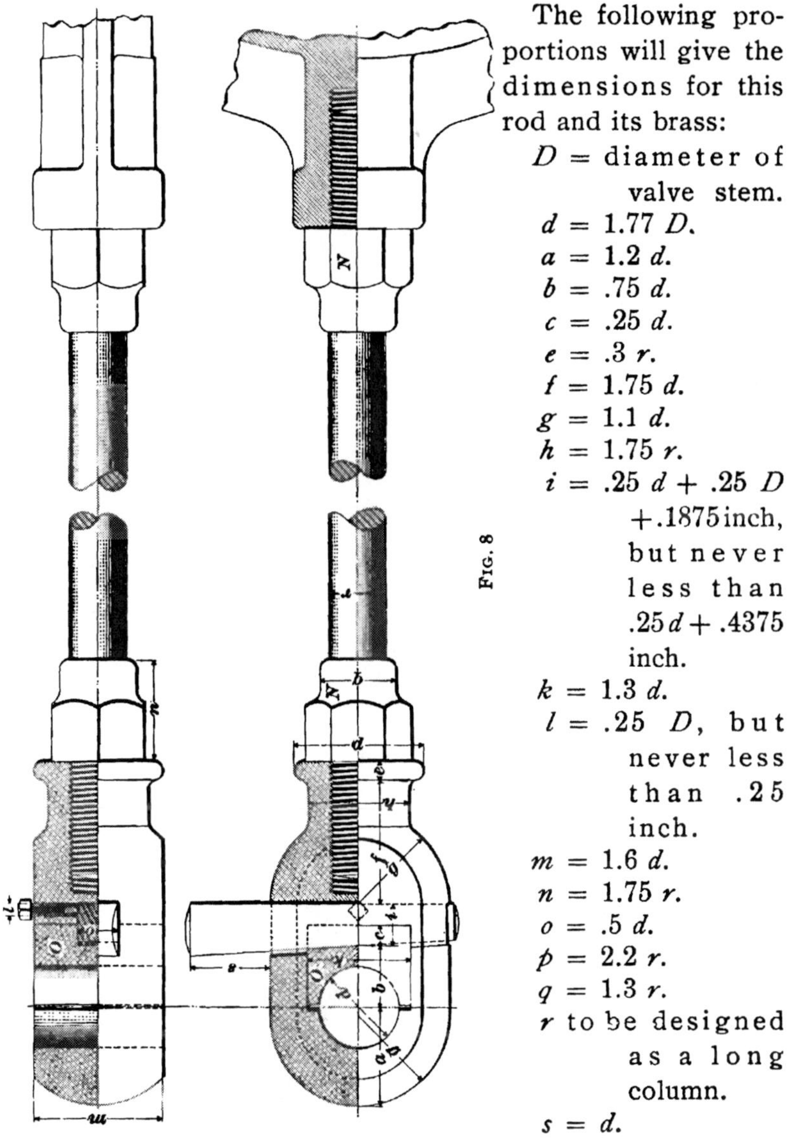

Fig. 8

The following proportions will give the dimensions for this rod and its brass:

D = diameter of valve stem.
$d = 1.77\ D$.
$a = 1.2\ d$.
$b = .75\ d$.
$c = .25\ d$.
$e = .3\ r$.
$f = 1.75\ d$.
$g = 1.1\ d$.
$h = 1.75\ r$.
$i = .25\ d + .25\ D + .1875$ inch, but never less than $.25 d + .4375$ inch.
$k = 1.3\ d$.
$l = .25\ D$, but never less than .25 inch.
$m = 1.6\ d$.
$n = 1.75\ r$.
$o = .5\ d$.
$p = 2.2\ r$.
$q = 1.3\ r$.
r to be designed as a long column.
$s = d$.

Taper of cotter = $\frac{3}{4}$ inch per foot.

17. Fig. 9 shows an eccentric rod with a modification of the marine connecting-rod end for the valve-stem pin bearing.

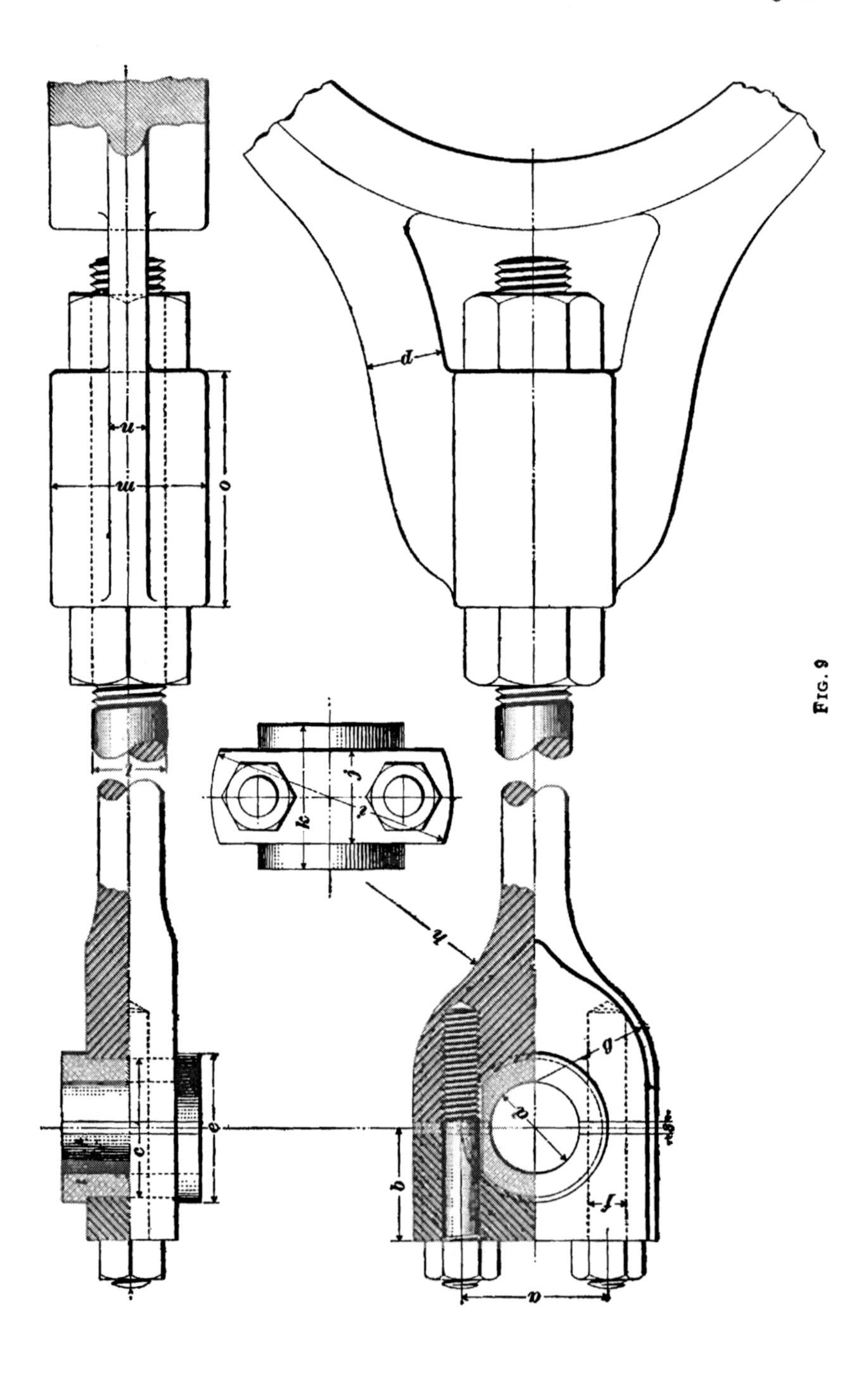

FIG. 9

The rod passes through a boss on the eccentric strap, and is fastened by the two nuts. This construction permits the valve to be adjusted when necessary. The bearing for the valve-stem pin is composed of brasses held in place by a wrought-iron cap and stud bolts. Liners are placed between the end of the rod and the cap, and the brasses may be adjusted by filing.

The proportions for this rod and for the boss that fastens it to the eccentric strap are as follows:

D = diameter of valve stem.
$d = 1.77\,D$.
$a = d + f + .375$ inch.
$b = 1.25\,d$.
$c = 1.5\,d$.
$e = 1.5\,d + .1875$ inch.
f = area at root of thread $= .38\,D^2$.
$g = .5\,i$.
$h = 1.5\,d$.
$i = 3.5\,f + d + .375$ inch.
$j = d$.
$k = 1.6\,d$.
l = diameter of eccentric rod at eccentric end.
$m = 2.1\,l$.
$n = .75\,D$.
$o = 3\,l$.
$p = 1.5\,D$.
$s = .125\,d$.

The diameter of the rod is determined by treating it as a long column.

ECCENTRIC SHEAVES AND STRAPS

18. An eccentric especially adapted for vertical engines is shown in Fig. 10; it has a cast-iron sheave and a steel strap. The sheave is made in two parts, so that it can be put on or removed from the shaft without disturbing flywheels or bearings. This is sometimes necessary, owing to the construction of the shaft, which will not permit an eccentric to be slipped into place over the end. The end of the eccentric rod is forged **T**-shaped and is fastened to the strap by tap bolts.

The two halves of the strap are held apart by liners, which permit of adjustment for wear. Split pins are put through

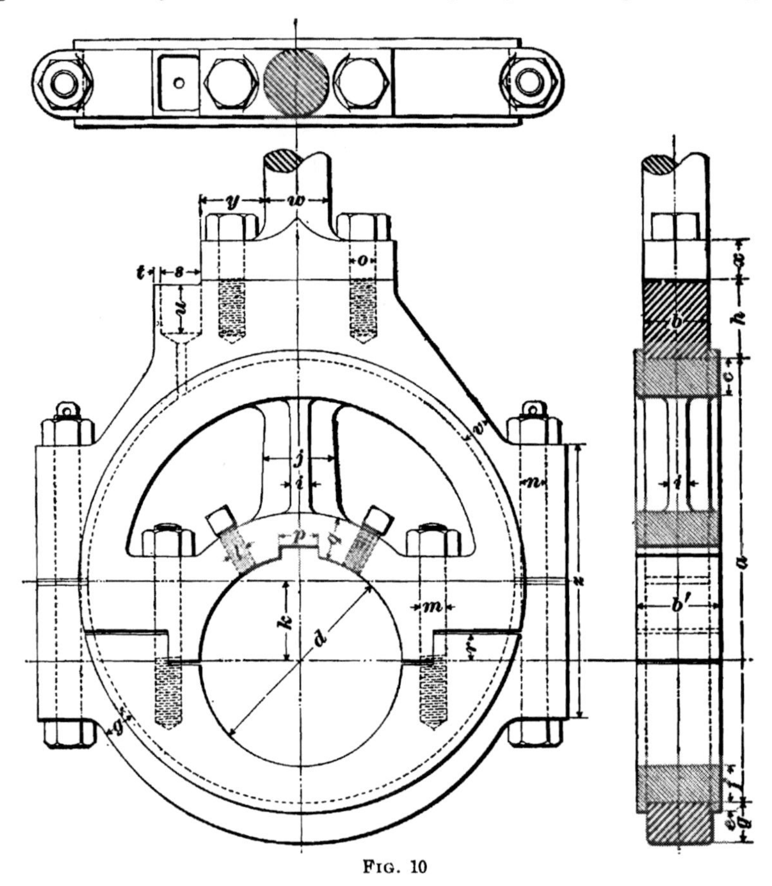

Fig. 10

the holes in the ends of the bolts n, to keep the nuts from turning off.

The following proportions give the necessary dimensions for this eccentric:

D = diameter of valve stem.

d = diameter of shaft.

$a = d + 2k + 2f$, never less.

$b = 2.5\,n$ at least, but never less than w.

$b' = b + .5\,c$.

$c = .6\,D + .375$ inch.

$e = .25\,c$.

$f = .7\,D + .5$ inch, unless more is required to allow nuts to be placed on stud *m*.

$g = .6\,D + .375$ inch.

$g' = .5\,D + .25$ inch.

$h = 3\,o$.

$i = .5\,c$.

$j = 2\,c$.

$k =$ eccentricity.

$l = .75$ of diameter of bolt *m*.

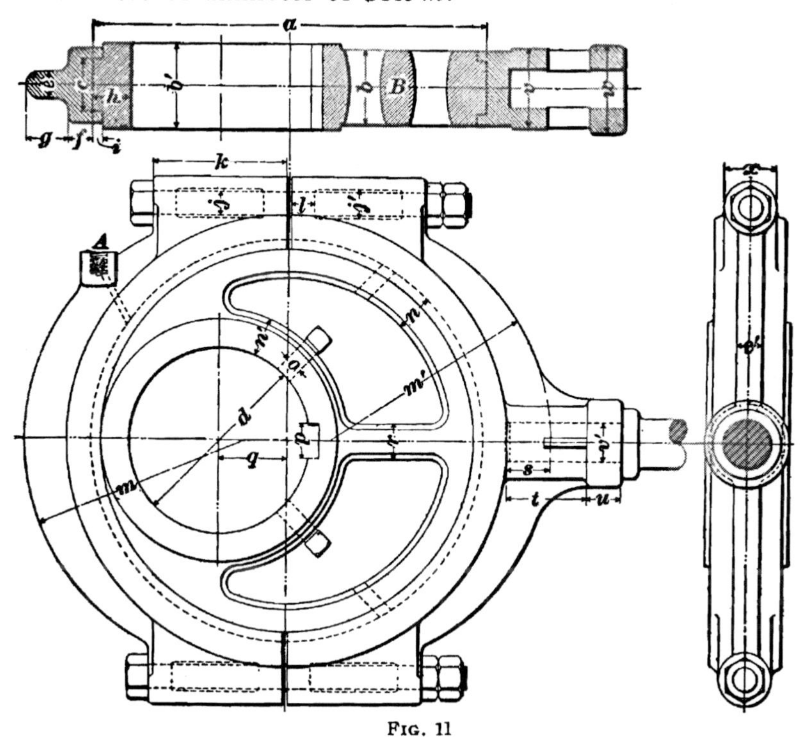

Fig. 11

Area of bolt *m* at root of thread $= .38\,D^2$; use nearest standard size of bolt.

$n =$ diameter of bolt *m*.

$o =$ diameter of bolt *m*.

$p = .7\,D + .5$ inch.

$q = .7\,D + .5$ inch.

$r = .125\,d$.

$s = D$.

t = .25 inch, constant.
u = 1.25 D.
v = .5 D + .25 inch.
w = diameter of eccentric rod.
x = .6 w.
y = 2.5 o.

z is to be found by laying out. The bolt n should clear the eccentric sheave by $\frac{1}{4}$ inch on all sizes up to D = $1\frac{1}{2}$ inches; $\frac{3}{8}$ inch for sizes of D from $1\frac{1}{2}$ to 2 inches; and $\frac{1}{2}$ inch for all sizes above D = 2 inches.

19. In Fig. 11, both the eccentric sheave and the strap are made of cast iron. The eccentric sheave is cast solid, and must therefore be slipped over the end of the shaft. The eccentric rod is held in a boss on the strap by means of a cotter.

It will be seen that in this case the strap is grooved for the sheave, while in Fig. 10 the groove is in the sheave. The construction that places the groove in the strap has the advantage of retaining oil better.

For eccentrics used with valve stems $\frac{1}{2}$ inch or less in diameter, the holes for bolts j are not to be cored.

A shows the boss for the oil cup, and B the cross-section of rib r.

The proportions are:

D = diameter of valve stem.
d = diameter of shaft.
a = $d + 2q + 2h$.
b = $2D$ + .125 inch.
b' = $2.25D$ + .125 inch.
c = 1.5 D.
e = .75 D.
e' = .75 D.
f = .7 D.
g = 1.25 D.
h = D + .125 inch.
i = .25 D + .0625 inch.
j = area of bolt at root of thread = .38 D^2; use the nearest standard size bolt.
j' = j + .1875 inch.
k = 4 D.
l = j.
$m = \frac{d + 2q + 2h + 2f}{2}$.
m' = m.
n = D + .125 inch.
n' = D + .125 inch.
o = .75 j.

$p = D.$

$q =$ eccentricity.

$r = D.$

$s = 1.25\,D.$

$t = 2.25\,D + 1.25$ inch.

$u = D.$

$v = 2.25\,D.$

$v' = 1.125\,D.$

$w = 2.5\,D.$

$x = 2.25\,j.$

STUFFINGBOXES

20. A **stuffingbox** of the ordinary form is shown in Fig. 12. The gland may be made of brass, of cast iron lined with brass, or simply of cast iron. The brass lining, however, injures the rod less than the harder iron. The gland is usually held in place by two stud bolts, but for

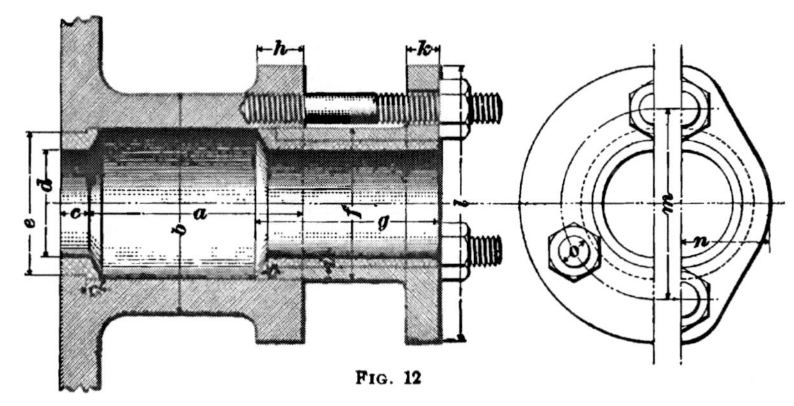

Fig. 12

large rods the gland is sometimes made circular instead of oval, and fastened by three or more studs.

The proportions for the gland shown in Fig. 12 are:

$d =$ diameter of rod.

$a = 1.6\,d + 1.5$ inches.

$b = 1.75\,d + 1.125$ inches.

$c = .1\,d + .75$ inch.

$c' = .5\,c.$

$e = 1.25\,d + .375$ inch.

$f = 1.25\,d + .625$ inch.

$g = 1.5\,d + 1$ inch.

$h = .3\,d + .5$ inch.

$i = .04\,d + .1875$ inch.

$k = .25\,d + .25$ inch.

$l = 2.25\,d + 1.75$ inches.

$m = 1.6\,d + 1.25$ inches.

$n = .75\,d + .375$ inch.

$o = .25\,d + .25$ inch for two bolts.

$= .2\,d + .25$ inch for three bolts.

$= .05\,d + 1.0625$ inches for four bolts.

$t = i.$

Two bolts should be used for glands on rods up to 3.5 inches in diameter. Above that size the gland should be made round, while three bolts should be used for rods up to 5.5 inches in diameter, and four bolts for all larger sizes.

21. For very high steam pressures, various styles of metallic piston-rod packing are used, one form of which is shown in Fig. 13. The construction of the stuffingbox and gland for this packing is very similar to the form shown in Fig. 12, but the packing is made up of rings of brass or similar antifriction metal. These rings are made in two or

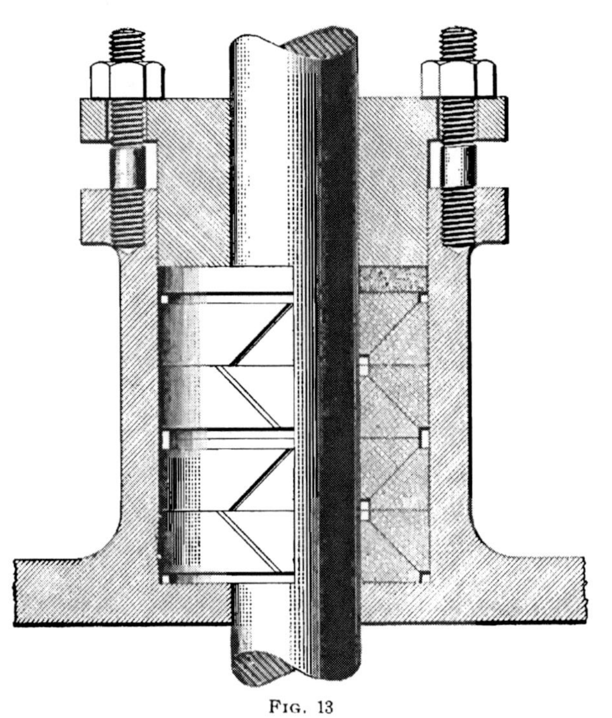

Fig. 13

more segments, depending on the size of the rod. By reason of the conical shape of the rings, the pressure of the gland forces the inner ones against the rod, while the outer ones are pressed against the sides of the stuffingbox. A rubber or fibrous ring placed between the gland and the first ring serves as a cushion to make the packing slightly elastic.

22. A stuffingbox of the form shown in Fig. 14 is generally used for small work, such as the spindles of valves, etc. The outside of the stuffingbox is threaded to receive a hexagonal nut, which fits over the gland. As the nut is screwed down, the gland is pressed downwards and compresses the packing.

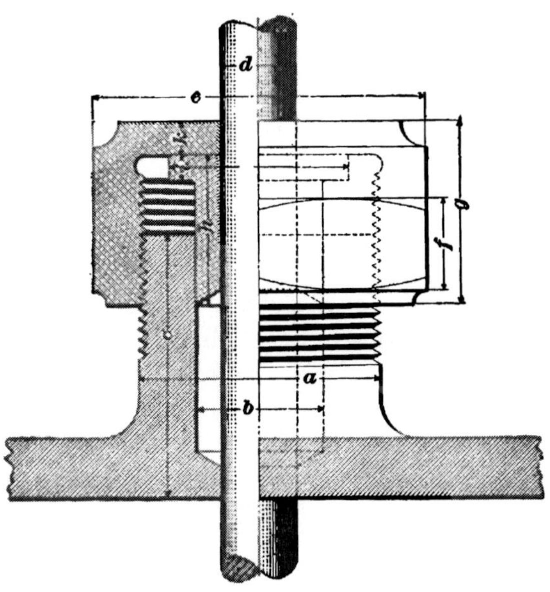

Fig. 14

The proportions used are:

d = diameter of rod.

$a = 2.5\,d + .5$ inch.

$b = 1.5\,d + .125$ inch.

$c = 3\,d + .25$ inch.

$e = 3.5\,d + .625$ inch.

$f = d + .125$ inch.

$g = 2\,d + .25$ inch.

$h = 1.5\,d + .25$ inch.

$i = .25\,d + .0625$ inch.

$k = .5\,d$.

This design may be used for rods up to $1\frac{1}{4}$ inches in diameter.

The number of threads per inch should be made the same as for a bolt having a diameter equal to the diameter of the rod.

ENGINE FLYWHEELS

23. Flywheels are subjected to a variety of complicated stresses, and it is therefore impossible to base their design on theory alone. Empirical rules representing successful practice are also an unsafe guide outside of the range of practice from which they are deduced, and, further, this range is generally unknown to the designer. The most satisfactory method is to work out the required weight of the flywheel and some of the principal dimensions rationally, and then calculate the detail dimensions by empirical rules deduced from practice. After this, the most important stresses may be calculated rationally, and, finally, if the last computation shows that the factor of safety is too small, such modifications should be made as will increase the factor to the necessary amount.

Small flywheels are generally one solid casting, the limit to the size of this construction being about 10 feet in diameter; but, even in this small size, the hub is often split so as to relieve cooling stresses in the casting. The two halves of the hub must then be given some mechanical connection. Wheels from 10 to 15 feet in diameter are generally cast in halves, while still larger wheels are usually cast in several pieces, the hub being cast separate and in halves, the arms separate, and the rim in at least two segments. These pieces are usually connected by bolts and links.

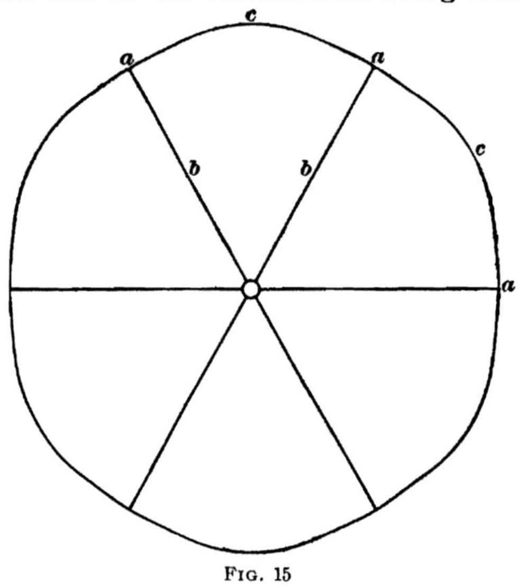

FIG. 15

24. Distortion of Flywheel Rim.—A thin ring rotating about its own center would be subject to the action of centrifugal force, which would act outwards uniformly all over the ring. Under this section, the ring, if composed of

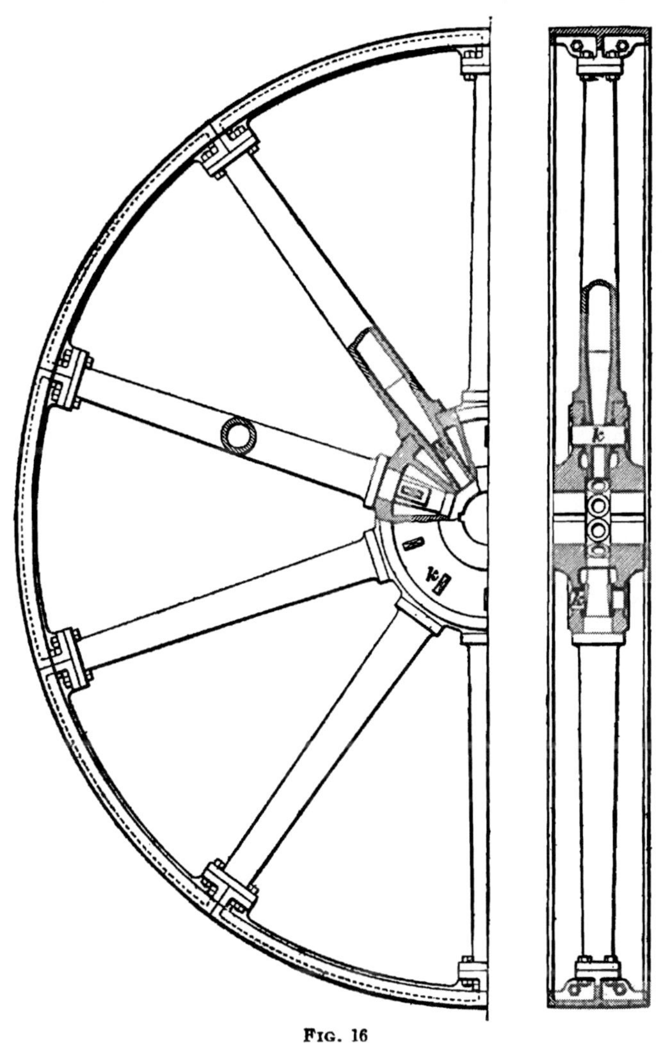

Fig. 16

elastic material, would expand. Now, if, at certain points, the ring should be held to its original diameter, as by the arms of a flywheel, it would take the distorted form shown somewhat exaggerated in Fig. 15. The rim is restrained at the points a, a by the arms b, b, and is bent outwards at c, c by centrifugal force. In the actual flywheel, the action is modified by the fact that the arms themselves stretch outwards under their own centrifugal force and the pull exerted on them by the rim, but the general effect remains as shown.

It is apparent from these remarks that a rotating flywheel with a small number of heavy arms will suffer greater distortion than one with a large number of light arms having the same total strength. In other words, the farther apart the arms, the greater the distortion of the rim between the arms.

25. A common form of construction for large flywheels is shown in Fig. 16. Particular attention is here called to the manner of attaching the segments of the rim to each other and to the arms. This method is superior to that in

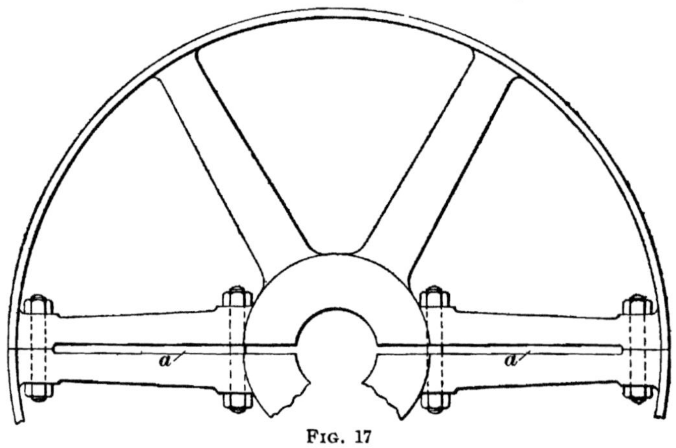

FIG. 17

which the segments of the rim are bolted together half way between the arms. The attachments of the rim to the arms as shown in the figure do not weaken the rim, but, on the contrary, tend to stiffen it and make the joint at the arms as strong as the solid rim. Furthermore, it is apparent from Fig. 15 that the distortion of a rotating wheel tends to open a joint half way between the arms, while it tends to close a radial joint directly over the arm.

If the joint in the rim cannot be placed directly over the arm, the best place for it is at a distance from the middle of the arm equal to .211 of the distance between middle lines of adjacent arms, measured along the middle of the rim. This value was determined by regarding the section of the rim from one arm to the next as a beam fixed at the ends and uniformly loaded, and finding, by means of higher mathematics, the point where the bending moment is zero.

26. Wheels in halves should have double arms along the line of separation, as shown at *a*, *a*, Fig. 17, especially if the rims are thin. The spokes, or arms, of flywheels are generally made of elliptical cross-section. In good practice, the major axis of the ellipse is placed in the plane of rotation, and is usually made from two to three times the length of the minor axis. For a high-speed wheel, an arm in which the minor axis is short has the advantage over one in which it is long, as the thinner and more wedge-shaped sections have much less air resistance. This is an important feature in a wheel traveling at a rim speed of about a mile a minute. The longer major axis, which is used with the shorter minor axis, also gives the arms greater resistance to the bending actions to which they are subjected through belt pull or variations in the speed of the wheel. The arms generally taper from the hub to the rim. The maximum taper found in good practice is such that the cross-sectional area at the rim is two-thirds of that at the hub.

27. Tension in Flywheel Rim.—In the design of a flywheel, the first value to be determined is the linear velocity at which the rim is to be run.

Let v = linear velocity of center of gravity of rim section, in feet per second; this center of gravity may be taken with sufficient accuracy as being at the mean rim radius from the center of the shaft;

S = stress per unit area of cross-section of rim;

H = weight of material in rim, in pounds per cubic foot.

The stress in the flywheel rim depends on the weight per cubic foot of the material in the rim and on the velocity of the mean rim section. From the principles of mechanics and strength of materials, the unit stress in the flywheel rim due to centrifugal force is expressed by the formula

$$S = \frac{H v^2}{4,631} \qquad (1)$$

In the case of cast iron, $H = 450$, and formula **1** reduces to

$$S = .09717\, v^2 \qquad (2)$$

For convenience in making calculations, formula **2** is frequently expressed by the approximate formula

$$S = .1\, v^2 \qquad (3)$$

Multiplying both sides of formula **1** by the area of cross-section of the rim in square inches, it becomes

$$T = S A = \frac{H A v^2}{4,631}, \qquad (4)$$

in which T = total tension in a cross-section of the rim, in pounds;

A = area of cross-section of rim, in square inches.

EXAMPLE.—What is the stress in the rim of a cast-iron flywheel when it has a velocity of 80 feet per second?

SOLUTION.—Apply formula **2**. Substituting $v = 80$, then
$S = .09717 \times 80 \times 80 = 621.9$ lb. per sq. in. Ans.

28. Cheapness of construction would always lead to building a wheel with a light rim of large diameter, rather than one with a heavier rim of smaller diameter. That is, high rim speed is desirable on the ground of economy or low first cost; but, according to the formulas of Art. **27**, the stress in the material increases as the square of this speed, and a limit is soon reached, beyond which it is not safe to go.

In good practice, the average velocity of flywheel rims, at least with wheels of moderate size cast in one piece, is about 70 feet per second, which, for cast iron, makes $S = 475$, nearly. Larger wheels cast in several pieces and bolted together are often run at a speed of about 88 feet per second, making $S = 750$, nearly. These limits of speed cannot be safely exceeded with cast-iron wheels of ordinary construction. If higher speeds are necessary, material of greater tensile strength in proportion to its weight must be employed for the rim, and special constructions must be used in the arms and hub.

29. Size of Flywheel Rim.—If the value of v is decided on, the weight of the wheel can be calculated. It is customary to base the calculations on the supposition that the entire weight is in the rim. The weight of the arms and hub is then additional. This results in making the wheel regulate the speed somewhat closer than was calculated, which is an advantage. The required weight of the rim for any given case is found as shown in *Mechanics of the Steam Engine*.

Let D = mean diameter of rim in feet $= \frac{60\,v}{\pi\,N}$;
v = velocity of rim, in feet per second;
N = number of revolutions per minute;
π = 3.1416;
W = weight of rim, in pounds;
h = weight of rim material per cubic inch, which is .261 for cast iron;
A = area of cross-section of rim, in square inches.

Then, the volume of the rim in cubic feet is $\frac{\pi\,D\,A}{144}$, and the weight of the rim is 1,728 h times this quantity, which gives the formula:

$$W = 12\,\pi\,D\,A\,h \qquad (1)$$

or,

$$A = \frac{W}{12\,\pi\,D\,h} \qquad (2)$$

Example.—What is the area of cross-section of rim of a flywheel having a mean diameter of 12 feet, if the rim is required to weigh 5,288 pounds?

Solution.—Applying formula **2**, the cross-section of the rim is

$$A = \frac{5{,}288}{12 \times 3.1416 \times 12 \times .261} = 44.8 \text{ sq. in.} \quad \text{Ans.}$$

30. If the wheel is not to carry a belt, the rim will probably be approximately rectangular in cross-section, the depth being usually from 1.1 to 1.4 times the breadth. The usual form of the rim for such a wheel is shown in Fig. 18. This illustration also shows the general construction for a small wheel to be run at low velocity; the whole wheel is one casting. In this construction, cooling stresses from casting are sure to occur, and are likely to exceed in magnitude the stresses that can be calculated. Also, the interior of a heavy rim is liable to have porous spots and other defects that cannot be detected. The only way to provide against accident due to these defects in a wheel of this kind is to keep down its rim speed, which should therefore never

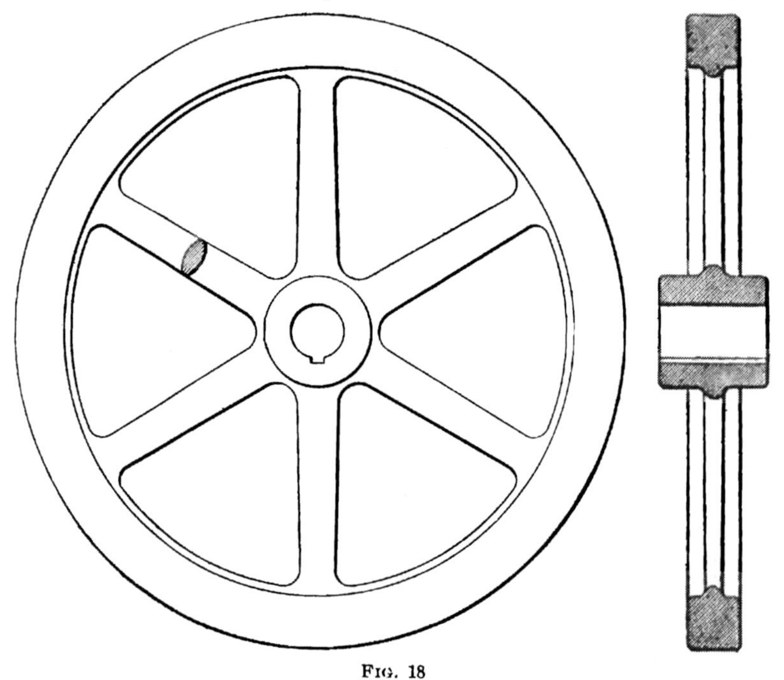

FIG. 18

exceed a limit of 70 feet per second, as stated in Art. **28.** Further calculation of stresses in the rim of a wheel of this construction is practically useless.

31. If a belt is to run on the flywheel, the width of the wheel must be designed to suit the belt.

Let D' = outer diameter of flywheel, in feet;

D = mean diameter of rim, in feet;

d = thickness of belt, in inches; for convenience, d will be taken as $\frac{1}{4}$ inch for a single belt, and $\frac{1}{2}$ inch for a double belt, although the actual thicknesses may vary slightly from these values;

f = coefficient of friction between belt and wheel, which may be taken as .25 in a design of this kind;

N = number of revolutions of the engine per minute;

P = maximum allowable tension on belt, in pounds per square inch of cross-section, which is from 250 to 400, with 300 as an average value;

T_1 = tension on tight side of belt, in pounds;

T_2 = tension on loose side of belt, in pounds;

V = velocity of outside of rim, in feet per minute;

t = thickness of flywheel rim, in inches, so that $D' = D + t$;

w = width of flywheel rim, in inches;

w_1 = width of belt, in inches;

h_1 = weight of belt in pounds per cubic inch, which may be taken as .035;

a = arc of contact between belt and wheel rim, in degrees;

H. P. = horsepower of engine.

It is first necessary to assume a value of t. This is done according to the best judgment of the designer, and at once gives the value of D', as D has already been found.

The difference in the tension on the two sides of the belt is $T_1 - T_2$, the belt pull, in pounds. As the whole of the horsepower of the engine is transmitted through this belt, the pull is also equal to the number of foot-pounds transmitted, 33,000 × H. P., divided by the number of feet passed through, $\pi D' N$. Hence, these two expressions for belt pull give the formula

$$T_1 - T_2 = \frac{33{,}000 \times \text{H. P.}}{\pi D' N} \qquad (1)$$

In order to determine the cross-sectional area of the belt required to transmit a given horsepower, it is necessary to ascertain the maximum tension T_1. As formula **1** gives only the difference $T_1 - T_2$, it is necessary to use another formula in connection with formula **1**. This formula, which is derived by means of higher mathematics, is

$$\log \frac{T_1}{T_2} = 2.729 f (1 - z) \frac{a}{360}, \qquad (2)$$

in which z is a factor depending on the centrifugal force of the belt. The other factors are as already given.

By means of higher mathematics, the value of z is found to be expressed by the formula

$$z = \frac{h_1 V^2}{9{,}660\ P} \qquad (3)$$

Then, for a velocity of 70 feet per second, when $P = 300$, the value of z is .213, and for 88 feet per second, $z = .337$, nearly.

By using formulas **1** and **2**, the value of T_1 may be found. Then, the width of the belt is found by the formula

$$w_1 = \frac{T_1}{d P} \qquad (4)$$

The rim of the flywheel is usually made from ½ to 1 inch wider than the belt. If the product of w as thus found and that of t as previously assumed is not sufficiently near the area of the cross-section of the rim as found by Art. **29**, a new value of t must be assumed, and the process repeated until the result is satisfactory.

EXAMPLE.—A flywheel having an average rim diameter of 12 feet and a rim section of 44.8 square inches is to be used on an engine of 200 horsepower at full load. The engine runs at 120 revolutions per minute, and the angle of contact is 180°. Find the dimensions of a flywheel rim of rectangular section to carry a double belt.

SOLUTION.—Assume that $t = 4$ in., or $\frac{1}{3}$ ft. Then, $D' = 12 + \frac{1}{3} = 12.33$ ft., and the velocity V becomes

$$\pi D' N = 3.1416 \times 12\tfrac{1}{3} \times 120 = 4{,}650 \text{ ft. per min.}$$

Then, by using formula **3**, making $h_1 = .035$, $V^2 = 4{,}650 \times 4{,}650$, and $P = 300$,

$$z = \frac{.035 \times 4{,}650 \times 4{,}650}{9{,}660 \times 300} = .261$$

Applying formula **2**, making $f = .25$, $z = .261$, and $a = 180$,

$$\log \frac{T_1}{T_2} = 2.729 \times .25\ (1 - .261) \frac{180}{360} = .25209,$$

and the number whose log is .25209 is 1.787.

Hence, $$\frac{T_1}{T_2} = 1.787 \text{ and } T_2 = \frac{T_1}{1.787}$$

Applying formula **1**, making H. P. = 200, and $\pi D' N = 4{,}650$,

$$T_1 - \frac{T_1}{1.787} = \frac{33{,}000 \times 200}{4{,}650}$$

Hence, $$T_1 = 3{,}223 \text{ lb.}$$

To find the width, apply formula **4**, making $T_1 = 3{,}223$, $d = .5$, and $P = 300$. Thus,

$$w_1 = \frac{3{,}223}{.5 \times 300} = 21\tfrac{1}{2} \text{ in., nearly.}$$

Hence, a 22-in. double belt should be used.

Then, $w = 22.5$ in., at least, and the area of the cross-section $= 4 \times 22\frac{1}{2} = 90$ sq. in., instead of the 44.8 sq. in. necessary. It is seen, therefore, that a 4-in. rim is about twice as thick as it should be. Therefore, try $t = 2$ in. Going through the calculations the same as before, w_1 now becomes 21.6. As before, a 22-in. belt will be used.

To get 44.8 sq. in.,

$$w = \frac{44.8}{2} = 22.4 \text{ in. Ans.}$$

This may be increased to 22.5 in., and used with a 22-in. belt.

32. Flywheel Arms.—The dimensions of the arms, or spokes, of the flywheel may now be calculated. The turning of the shaft by the engine is resisted by the belt pull and the inertia of the flywheel rim, and this produces a bending stress in the arms. Each arm may then be considered as a cantilever loaded at one end. Then, by finding expressions for the bending moment and for the resisting moment of the section of the arm nearest the hub, and placing them equal to each other, a formula for the size of arm is derived.

Let S_4 = safe bending stress in outer fiber of arm, at hub, in pounds per square inch; S_4 should not exceed 1,000 to 1,400 for cast-iron arms;

N = number of revolutions of flywheel per minute;

n = number of arms in wheel;

T = mean twisting moment, in inch-pounds, transmitted by shaft in ordinary driving, which tends to bend the arms about the hub;

I = moment of inertia of cross-section of arms, dimensions in inches;

c = distance from neutral axis to outside edge, in inches.

The twisting moment of the flywheel is equal to the total bending moment on all the arms and also to the product of the difference between the belt tensions, or the equivalent from the power transmitted, and the radius of the flywheel in inches. It would be correct to use the length of the arms, but the radius is more convenient and its use has the effect of increasing the factor of safety. Hence, multiplying formula **1**, Art. **31**, by the outside radius in inches gives the formula for T as

$$T = \frac{198{,}000 \text{ H. P.}}{\pi N} \qquad (1)$$

In a flywheel having a heavy rim, it is reasonable to assume that the turning moment is divided equally among the arms. Hence, the bending moment on each arm is $T \div n$, regarding the arm as a beam fixed at one end and loaded at the other. Placing this equal to the resisting moment of the arm,

$$\left.\begin{aligned} \frac{T}{n} &= S_4 \frac{I}{c} \\ \text{or,} \qquad \frac{I}{c} &= \frac{T}{n S_4} \end{aligned}\right\} \qquad (2)$$

Let A = outside depth, that is, the dimension in the direction of motion, of arm at hub, in inches;
a = inside depth of arm, if hollow, at hub, in inches;
B = outside breadth, that is, the dimension perpendicular to the plane of rotation, of arm, at hub, in inches;
b = inside breadth of arm, if hollow, at hub, in inches.

For some of the simple forms, the values of $\frac{I}{c}$ are:

For a solid rectangle, $\frac{I}{c} = \frac{A^2 B}{6}$.

For a hollow rectangle, $\frac{I}{c} = \frac{1}{6} \frac{(B A^3 - b a^3)}{A}$.

For a solid ellipse, $\frac{I}{c} = \frac{\pi}{32} B A^2$.

For a hollow ellipse, $\frac{I}{c} = \frac{\pi}{32} \frac{(A^3 B - a^3 b)}{A}$.

A may be taken as from $2B$ to $3B$, while suitable relations must be assumed between the interior and exterior dimensions of the cross-section, if the arm is hollow. The dimensions of the cross-section may then be determined by use of formulas **1** and **2**.

EXAMPLE.—A flywheel makes 120 revolutions per minute in transmitting 200 horsepower. If the power is transmitted through a flywheel with eight solid elliptical arms, with the major axis $2\frac{1}{2}$ times the minor axis, what are the dimensions of the arms at the hub?

SOLUTION.—Applying formula **1**, making H. P. = 200, $N = 120$, and $\pi = 3.1416$,

$$T = \frac{198{,}000 \times 200}{3.1416 \times 120} = 105{,}042$$

When $A = 2\frac{1}{2}B$, the value of $\frac{I}{c}$ is

$$\frac{\pi}{32}BA^2 = \frac{\pi \times 6.25\,B^3}{32} = .6136\,B^3$$

Then, applying formula **2**, making $\frac{I}{c} = .6136\,B^3$, $S_4 = 1{,}000$, $n = 8$, and $T = 105{,}042$,

$$.6136\,B^3 = \frac{105{,}042}{8 \times 1{,}000} = 13.13$$

Hence, $$B = \sqrt[3]{\frac{13.13}{.6136}} = 2.776 \text{ in.}$$

Then, $A = 2\frac{1}{2}B = 2\frac{1}{2} \times 2.776 = 6.94$, say 7, in.

Hence, at the hub, the arms would probably be $2\frac{7}{8}$, or 3 in. × 7 in. Ans.

33. Check Calculations for Built-Up Flywheels. In the foregoing calculations, the effect of the mutual pull of rim and spokes on each other has been entirely neglected. This pull, however, produces serious stresses, and to compensate for the neglect of these, the calculated stresses have been made very low. In the case of flywheels, this method of designing has sometimes been unsatisfactory, and the bursting of large flywheels has not been uncommon.

Therefore, in any built-up flywheel, except where the design is practically a repetition of one that has already proved safe in actual service, the following stresses should be calculated: (1) The tensile stress in the inner fibers of the rim directly over the arms; (2) the tensile stress in the outer fibers of the rim, half way between the arms; (3) the tensile stress in the outer fibers of the arms at the hub; and (4) the tensile stress in the arm at the rim.

Though the calculation of these stresses is somewhat laborious, it should be done for any built-up wheel except those just stated. In deriving the following formulas for the principal stresses, the arms were assumed to be of uniform cross-section throughout. In practice, the arms are frequently tapered; but the formulas are probably a close enough approximation to the real values of the stresses with tapered arms, provided the taper does not exceed that stated in Art. **26.** However, this is another reason why the stresses as calculated should be kept very low, as compared with those allowed in other portions of the engine.

34. The following formulas are based on formulas for the stresses in a built-up flywheel as derived by Professor Gaetano Lanza. They are derived by means of higher mathematics from a careful analysis of all the stresses in the flywheel due to the centrifugal force of the rim and the restraining effect of the arms.

Let A = area of cross-section of rim, in square feet;
A_1 = area of cross-section of arm at hub, in square feet;
A_2 = area of cross-section of arm at rim, in square feet;
H = weight of material, in pounds per cubic foot;
g = 32.16;
I = moment of inertia of cross-section of rim or arm about its neutral axis, the dimensions of the section being taken in feet;
c = distance from neutral axis to outermost fiber of section, in feet;
n = number of arms in wheel;
R = distance from center of hub to middle of cross-section of rim, in feet;
r_1 = distance from center of hub to outer end of arm, in feet;
r_2 = radius of hub, in feet;
T = mean twisting moment transmitted by shaft, in foot-pounds;
v = linear velocity of middle of cross-section of rim, in feet per second;
y_1 = distance from neutral axis of rim to inside of rim, in feet;

y_2 = distance from neutral axis of rim to outside of rim, in feet;

a = half the angle between middle lines of two adjacent arms;

K = a constant.

NOTE.—If the cross-section of the rim is symmetrical about a line through its middle, drawn perpendicular to the plane of rotation of the wheel, this axis of symmetry is also the neutral axis.

The tensile stress T_1, due to bending, in the inner fibers of the rim over the arms, in pounds per square inch, is expressed by the formula

$$T_1 = \frac{Hv^2}{144g}\left\{1 + \frac{K}{6}\left[\frac{57.3\,R y_1}{I a} - \left(\frac{1}{A} + \frac{R y_1}{I}\right)\cot a\right]\right\} \quad (1)$$

The tensile stress T_2, due to bending, in the outer fibers of the rim, half way between arms, in pounds per square inch, is expressed by the formula

$$T_2 = \frac{Hv^2}{144g}\left\{1 + \frac{K}{6}\left[-\frac{57.3\,R y_2}{I a} + \left(\frac{R y_2}{I} - \frac{1}{A}\right)\operatorname{cosec} a\right]\right\} \quad (2)$$

The tensile stress T_3 in the outer fibers of the arms at the hub, in pounds per square inch, is expressed by the formula

$$T_3 = \frac{H v^2 K}{432 g A_1} + \frac{T c}{144\, n\, I} \quad (3)$$

The tensile stress T_4 in the arms at the rim, in pounds per square inch, is given by the formula

$$T_4 = \frac{H v^2 K}{432 g A_2} \quad (4)$$

The original formulas were derived for stresses in pounds per square foot, but the factor 144 has been introduced in the denominators of the right-hand members of the formulas, so that the stresses are now expressed in pounds per square inch and can be compared directly with other stresses. The units of length and area, however, are expressed in feet and square feet, and should be so substituted in applying the formulas.

The value of K is determined by the formula

$$K = \frac{3 - \left(\frac{r_1 - r_2}{R}\right)^2 \left(\frac{r_1 + \frac{1}{2} r_2}{R}\right)}{\frac{1}{A_1}\left(\frac{r_1 - r_2}{R}\right) + \frac{57.3}{2 A a}} \quad (5)$$

35. If the calculation of the stress by the foregoing methods does not show a large factor of safety, the design must be modified accordingly. This may be done by any one of the following methods or combinations of them:

1. By lowering the linear velocity of the rim. This is the most certain means of affecting the result. As the engine speed cannot be changed, the diameter of the wheel must be reduced. Changing the rim speed involves a complete repetition of all the work of design, beginning with a recalculation of the weight. A lowering of the rim speed also increases the cost of the wheel, as stated in Art. **28.**

2. If the speed has not been assumed at a higher value than 88 feet per second, as given in Art. **28,** it is better to seek increased safety by increasing the *strength* of the wheel, rather than by decreasing the speed. The strength may be increased by any of the following methods: Altering the disposition of the metal in the rim, so as to give the rim greater strength against bending in the plane of rotation. This is often done by putting on ribs extending inwards from the rim. When this is done, it should always be determined by calculation that the modulus of the cross-section, $\frac{I}{c}$, of the rim is really increased, as it is possible to add ribs in such a way as to weaken the rim. A sure means of accomplishing the desired effect is to decrease the width and increase the thickness of the rim, making it hollow if necessary; but if the wheel is to carry a belt, the extent to which the width can be reduced is limited by the width of the belt.

3. By using material of greater tensile strength in proportion to its weight. This method is often successfully employed. A large number of flywheels with wooden rims have been designed and built, and have also been very successful. It is now common to use a steel rim cast in at least two segments separate from the arms, the whole being bolted together in assembling the wheel. Rims are also sometimes built up of boiler plates.

4. By increasing the number and decreasing the size of the arms. Excellent results have been attained by this

method. Band-saw wheels, built like bicycle wheels, with small adjustable pipe arms 6 or 8 inches apart along the rim, are successfully run at 10,000 feet per minute. This is a special construction, and is too expensive for ordinary engine flywheels.

It should be borne in mind that nothing can be done toward increased safety by simply enlarging the dimensions of the rim without changing its shape, material, or construction. Any increase in the weight of the rim increases the centrifugal force that tends to tear the wheel apart, in just the same proportion that it increases the cross-section tending to resist rupture of the rim.

36. Flywheel-Rim Joints.—The rim joints may now be designed. Fig. 19 shows the usual form when the rim joint comes directly over an arm, the rim of the thickness t being broken off at r, r. There will usually be a line of bolts with center lines as at $e f$, and a line of bolts on each side of the joint with center lines as at $h i$ and $h_1 i_1$. At m is shown the outer end of an arm whose center line coincides with the part $d k$ between the two rim sections.

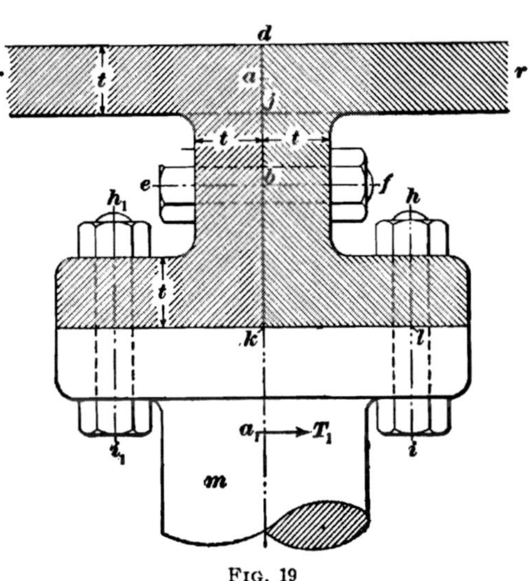

FIG. 19

Let a, Fig. 19, indicate the center of gravity of the cross-section of the rim, not including the flanges for the joint. The stresses in the rim due to bending produce an equivalent tension T_1, which is given by a formula also derived by Professor Lanza. Thus,

$$T_1 = \frac{H v^2}{g}\left(A - \frac{K}{6} \cot a\right) \qquad (1)$$

This tension acts at a point a_1, Fig. 19, such that $a a_1$ is given by the formula

$$a a_1 = \frac{KR\left(\frac{57.3}{a} - \cot a\right)}{6A - K \cot a} \quad (2)$$

The symbols are the same as those used in Art. **34.**

By taking moments about the point d, Fig. 19, the tension on the bolts at e, f, due to the tension in the rim, is $T_1 \times d a_1 \div d b$. To this must be added the tension on the bolts due to screwing up. This may be estimated as $T_2 = 18{,}700 \times d_1$, where d_1 is the diameter, in inches, of the bolts at the bottom of the threads. The cross-section at the bottom of the threads of the bolts through the joint at $e f$ must be sufficient to carry the total tensile stress on them, as just calculated. The formula for the diameter of bolts then becomes

$$d_1 = \frac{11{,}900}{S_1 n} + \sqrt{\frac{T_1 \times \overline{d a_1}}{.7854 \overline{d b} \times S_1 n} + \left(\frac{11{,}900}{S_1 n}\right)^2}, \quad (3)$$

in which S_1 = safe tensile stress in bolts, in pounds per square inch;

n = number of bolts in flange at distance $d b$ from outside of rim;

T_1 = equivalent tension, in pounds, acting at a_1;

$d a_1$ and $d b$ = distances on the rim, as shown in Fig. 19, measured in inches.

Ordinarily, the factor of safety in the bolts of the rim joints of existing flywheels is low.

Owing to close competition, designers have sometimes allowed stresses as high as 12,500 pounds per square inch of cross-section in wrought-iron bolts, and 15,000 in steel bolts, but these values are excessive; the stresses should be kept as low as possible.

37. Stress in Rim Flange.—The stress in the rim flange should also be calculated. This stress is composed of a combination of two stresses—that due to the tendency to bend at the point where the flange is attached to the rim, as at j, Fig. 19, owing to the equivalent tension T_1, and that due to the centrifugal tension, as calculated by formula **4**, Art. **34.**

The bending moment of T_1 about the point j is $T_1 \times j a_1$, and the moment of resistance of the section of the flange at j is $\frac{S w t^2}{6}$, which gives the stress due to bending as

$$S = \frac{6 T_1 \times \overline{j a_1}}{w t^2}, \qquad (1)$$

in which T_1 = total tension at a_1, as calculated by formula **1**, Art. **36;**

w = width of flange, in inches;

t = thickness of flange, in inches;

S = stress in flange due to bending.

Then, adding this stress to the tensile stress due to the centrifugal force, as found by formula **4**, Art. **34**, gives the total stress S_1 as

$$S_1 = \frac{6 T_1 \times \overline{j a_1}}{w t^2} + \frac{H v^2 K}{6 g w t} \qquad (2)$$

in which H, v, g, and K are as given in Art. **34**, and the other symbols are the same as just stated. It will be noticed that $w t$, in square inches, appears in the last part of the formula in place of the area $144 A_2$ of cross-section of the arm, and as there are two flanges of thickness t to take the stress due to centrifugal force, the denominator is $6 g w t$ instead of $432 g A_2$.

At first, the thickness of rim and flange should be made the same, to improve the chances of a good casting; then, S_1 should be calculated by the foregoing formula. If S_1 should appear excessively large, the thickness of the flange should be increased, until S_1 is reduced to a safe value.

38. Stress in Bolts Fastening Arm to Rim.—The bolts fastening the rim flange to the arms are located as shown at $h i$ and $h_1 i_1$, Fig. 19. The total stress in these bolts is the sum of the stress due to the centrifugal force, that due to screwing up the nuts, and that due to the tension T_1 in the rim caused by bending. T_1 tends to produce a rotation of the flange about the point d, and the moment is

$T_1 \times d\,a_1$. This moment divided by $k\,l$, the perpendicular distance from $h\,i$ to d, gives the tension on the bolts. The tension due to screwing up the nuts may be taken, as in Art. **36,** as $T_2 = 18{,}700\,d_1$. The tension due to the centrifugal force is $\frac{H\,v^2\,K}{6\,g}$, which is $72\,A_2$ times the value of the tensile stress, in pounds per square inch, as given in formula **4,** Art. **34,** there being two sections and the total tension on each being considered. Hence, the total tension on these bolts is found by taking the sum of these three tensions, as shown by the formula

$$T = \frac{T_1 \times d\,a_1}{k\,l} + \frac{H\,v^2\,K}{6\,g} + 18{,}700\,d_1 \qquad (1)$$

The total tension as thus found should equal the product of the area of cross-section of half the bolts on one arm and the safe stress. This gives a formula for the diameter that reduces to $d_1 =$

$$\frac{11{,}900}{S_1\,n} + \sqrt{\frac{T_1\,d\,a_1}{.7854\,k\,l \times S_1\,n} + \frac{H\,v^2\,K}{4.7124\,S_1\,n\,g} + \left(\frac{11{,}900}{S_1\,n}\right)^2}, \qquad (2)$$

in which S_1 may have values as given in Art. **36,** and n is the number of bolts fastening one section of the rim to an arm, or half the total number of bolts fastening the rim to one arm.

39. Strength of Joint Between Arms.—When a built-up flywheel is made so that the joint comes between the arms, it should be located at a distance from the center line of one arm equal to $.211\,x$, where x represents the distance between the middle lines of adjacent arms, measured along the middle of the rim.

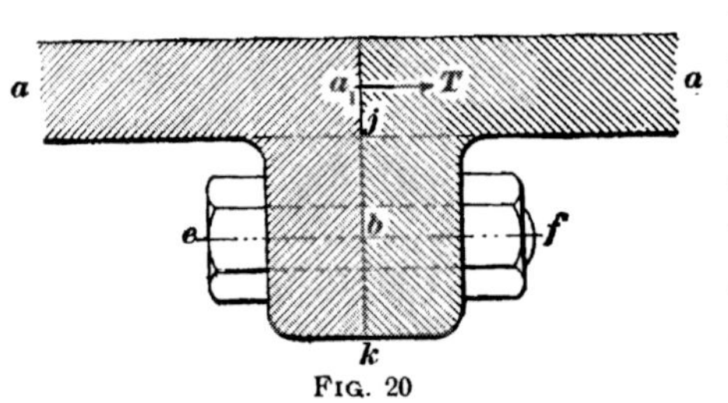

FIG. 20

Fig. 20 represents a joint located between arms as just stated. The rim is broken away at a, a, and the tension T in the rim acts at the point a_1, which may be taken, without serious error, as the center of gravity of the rim section, neglecting the flanges. The total tension on one joint located as just stated may be found by using the following formula, which has been derived by means of higher mathematics:

$$T = \frac{H v^2}{g}\left[A - \frac{K}{6}\,\frac{\cos\left(a - \frac{12.1\,x}{R}\right)}{\sin a}\right], \qquad (1)$$

in which the symbols are the same as explained in Art. **34.** R and x, of course, must be in the same units.

Then, as the tension of the bolts acts at b and the flange in opening would turn about the point k, the tension T_1 in the bolts may be found by using the formula

$$T_1 = \frac{T \times \overline{a_1 k}}{\overline{b k}} \qquad (2)$$

It is apparent that the stress T_1 in the bolts is diminished by putting them as close to the rim as possible; that is, by increasing $b\,k$. The tension T_2 due to screwing up may be estimated as being equal to 18,700 d, when d represents the diameter of the bolts in inches at the root of the thread. The cross-sectional area of the bolts at the bottom of the threads must be made sufficient to carry the total tension without excessive stress. Placing the product of the area of cross-section of a bolt, the number of bolts, and the stress equal to the sum of these tensions, gives the expression

$$.7854\, d^2 n S_1 = \frac{T \times \overline{a_1 k}}{\overline{b k}} + 18{,}700\, d$$

This reduces to the formula

$$d = \frac{11{,}900}{n S_1} + \sqrt{\frac{T \times \overline{a_1 k}}{.7854\, \overline{b k} \times S_1 n} + \left(\frac{11{,}900}{n S_1}\right)^2} \qquad (3)$$

40. Besides being subjected to tension, the bolts at e, f, Fig. 20, are subjected to a shearing force that may be calculated by the use of the following formula, which has been derived by means of higher mathematics:

$$S = \frac{H v^2 K}{6 g}\left[\frac{\sin\left(a - \frac{12.1\,x}{R}\right)}{\sin a}\right],$$

in which the symbols have the same meaning as in Art. **34.** R and x, of course, must be taken in the same units. The shear S is the total shear on all the bolts in one joint, and to get the unit stress, S must be divided by the area of cross-section of all the bolts in that joint. The shearing stress should not exceed 10,000 pounds per square inch for wrought iron nor 12,500 for steel, and, if possible, should be kept at a lower figure. The shearing stress is carried at the joint, so that the entire area of cross-section of the bolts, rather than that at root of threads, should be used here. The thickness of the flange may be made the same as that of the rim.

41. Flywheel Hub.—When the hub of a flywheel is cast in sections, it must be fastened together with fastenings designed to resist the forces that tend to rupture the hub. There are two common methods of securing together the parts of the hub—one is by shrinking wrought-iron or steel bands around the hub, and the other is by bolting. In the absence of more accurate information, it will be safe to assume that the force tending to separate the halves of the hub is equal to that which tends to tear the rim apart.

Consider first the method of fastening the hub by means of bands. From formula **4,** Art. **27,** the total force that tends to separate the rim at any section is $\frac{H A v^2}{4{,}631}$. If this is resisted by the rings, the area of the sections of the rings must be sufficient, so that the stress shall not be excessive. The product of the area of cross-section of one ring, the number of rings, and the stress must equal $\frac{H A v^2}{4{,}631}$; that is,

$$A_3 m S_1 = \frac{H A v^2}{4{,}631},$$

in which A_3 = area of section of one band, in square inches;
S_1 = safe tensile stress, in pounds per square inch;
m = number of bands, which will generally be two;
H = weight of metal in rim, in pounds per cubic foot;
A = area of cross-section of rim, in square inches;
v = velocity of rim, in feet per second;
g = 32.16.

Solving the foregoing expression for A_s gives the formula

$$A_s = \frac{H A v^2}{4{,}631\, m S_1} \qquad (1)$$

S_1 should not exceed from 1,500 to 3,000. The reason for limiting the calculated stress to this low value is that there is a much greater and uncalculated stress in the bands due to shrinking them on. In a cast-iron wheel, $H = 450$, and formula **1** becomes

$$A_s = .09717 \frac{A v^2}{m S_1} \qquad (2)$$

42. In a similar manner, if the sections of the hub are to be bolted together, the bolts must be designed to resist similar stresses. In fact, the formulas of Art. **41** may be used directly for the area of cross-section of the bolts in a joint on one side of the hub. In such a case, m would be the number of bolts on one side of the hub, and A_s the area of cross-section of one bolt.

In the case of bolts, however, there are no shrinkage stresses, but there is the stress due to screwing up the nut. The tensile stress in the bolts should not exceed 2,500 pounds per square inch of cross-section at the root of the threads.

The hub should be of good substantial dimensions, from about 2 inches thick in the smallest wheel to about 6 inches thick in the largest, and long enough to get a good solid bearing on the shaft. The hub contains an excess of metal for all calculable stresses on it, and it is not necessary to make any calculations regarding them.

43. General Construction of Flywheels.—As has already been stated, flywheels of small diameter are usually cast solid; the arms are of elliptical cross-section, and the wheel has the general appearance of a belt pulley with a heavy rectangular rim. Large flywheels, however, are cast in sections or built up of plates and castings.

A heavy flywheel with solid, elliptical arms is shown in Fig. 21. This wheel is cast in four sections, with two arms to each section, and the hub is formed of two separate rings, which are bolted and keyed to the segments forming the inner ends of each pair of arms. The segments of the rim are joined by means of steel or wrought-iron rings R, which are shrunk on bosses formed by recesses cast in the rim.

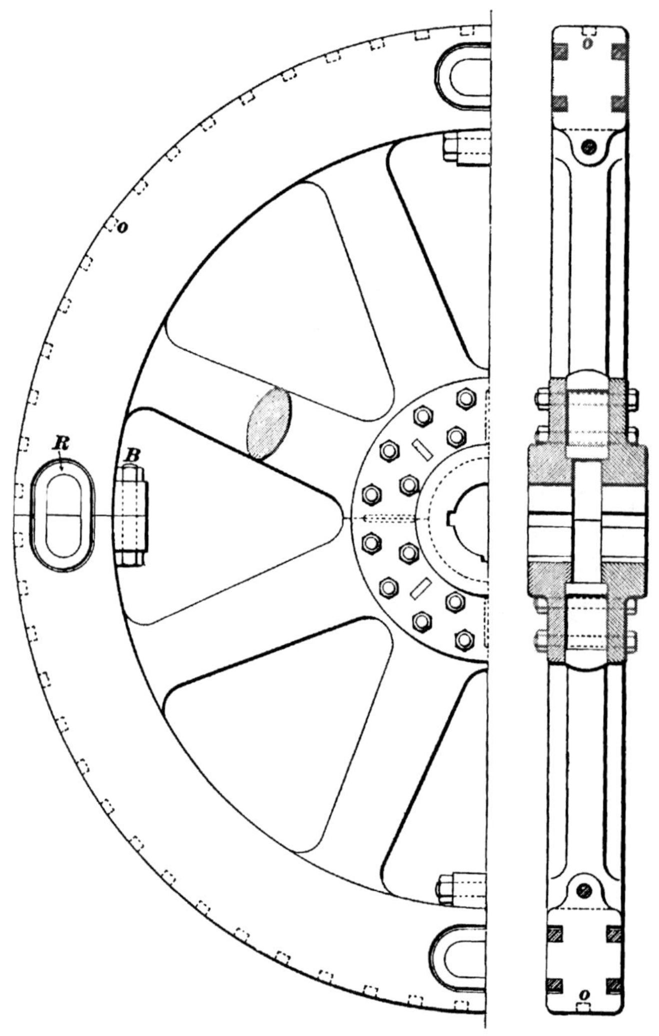

Fig. 21

Besides these rings, bolts B pass through lugs on the inner surface of the rim. In order that the wheel may be amply strong, the net section of the bolt and two rings must be sufficient to withstand the total tension due to centrifugal force tending to separate the rim through the section that they join. This tension may be calculated by formula **4**, Art. **27.**

The bolts and keys joining the inner ends of the arms to

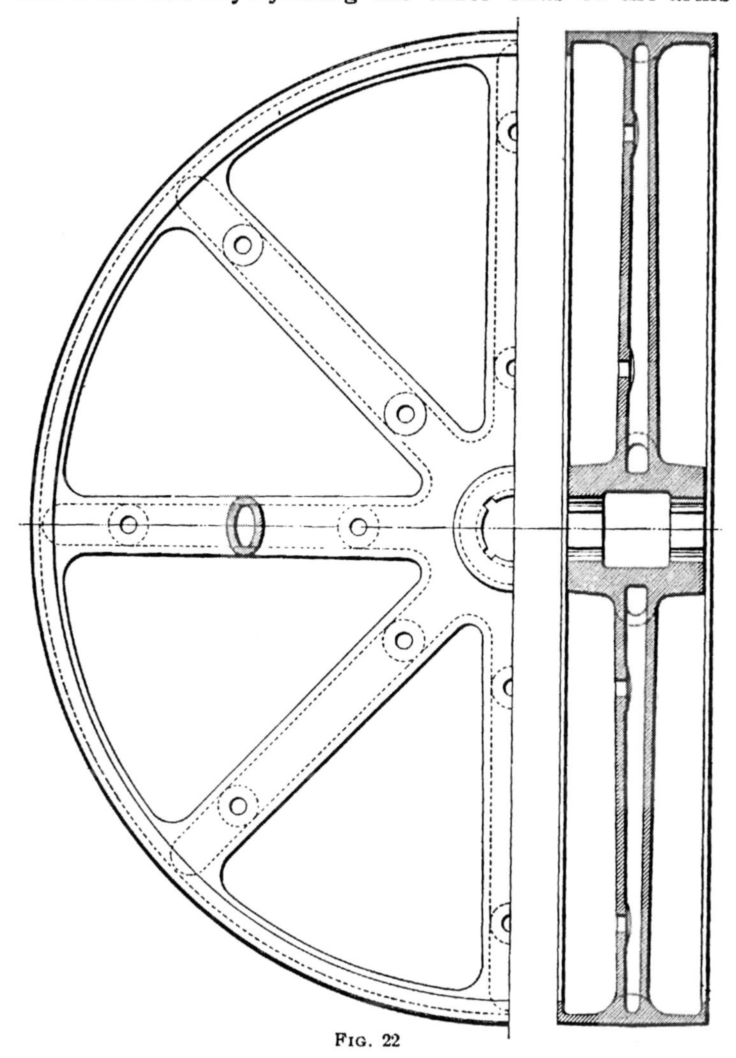

Fig. 22

the hub are in double shear, and they must be calculated to withstand the pull exerted by the rim on each arm. This pull, as derived from formula **4**, Art. **34**, is

$$F = \frac{H v^2 K}{3 g}$$

44. Fig. 22 shows a flywheel with the face of the rim turned to serve as a belt pulley. The arms are oval in section and cast hollow, thus giving them increased stiffness for a given weight. The rim is also given a channel-shaped section, which increases its ability to withstand the bending stresses produced by centrifugal force in the sections between the arms.

45. Another method of fastening the arms to the hub is shown in Fig. 23. Here, the ends of the arms are cast with

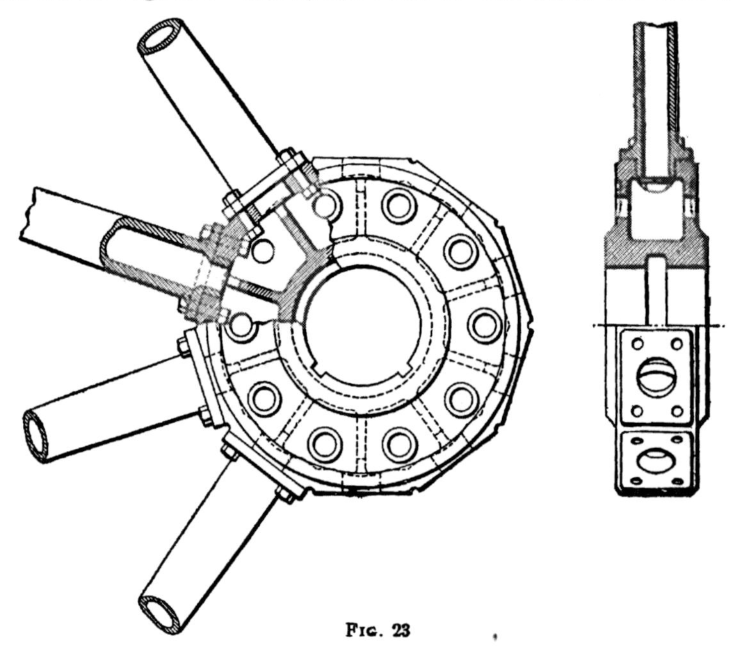

FIG. 23

flanges, and circular bosses, which fit into recesses bored in the hub, are also turned on them. Bolts pass through the flanges in the ends of the arms and in the face of the hub, thus holding the arms securely in place.

46. Fig. 24 shows a flywheel cast in halves. The sections of the rim are joined by means of steel or wrought-iron bars *b* inserted in holes cast in the ends of the rims. These bars are fastened to the rim by keys *k* that pass through holes fitted for that purpose. The arms are oval in section and are cast solid. The hub is provided with bosses through which four bolts are passed, thus joining the two parts of the hub securely. The holes *o* cored in the rim of the flywheels shown in Figs. 21 and 24 are for the purpose of inserting

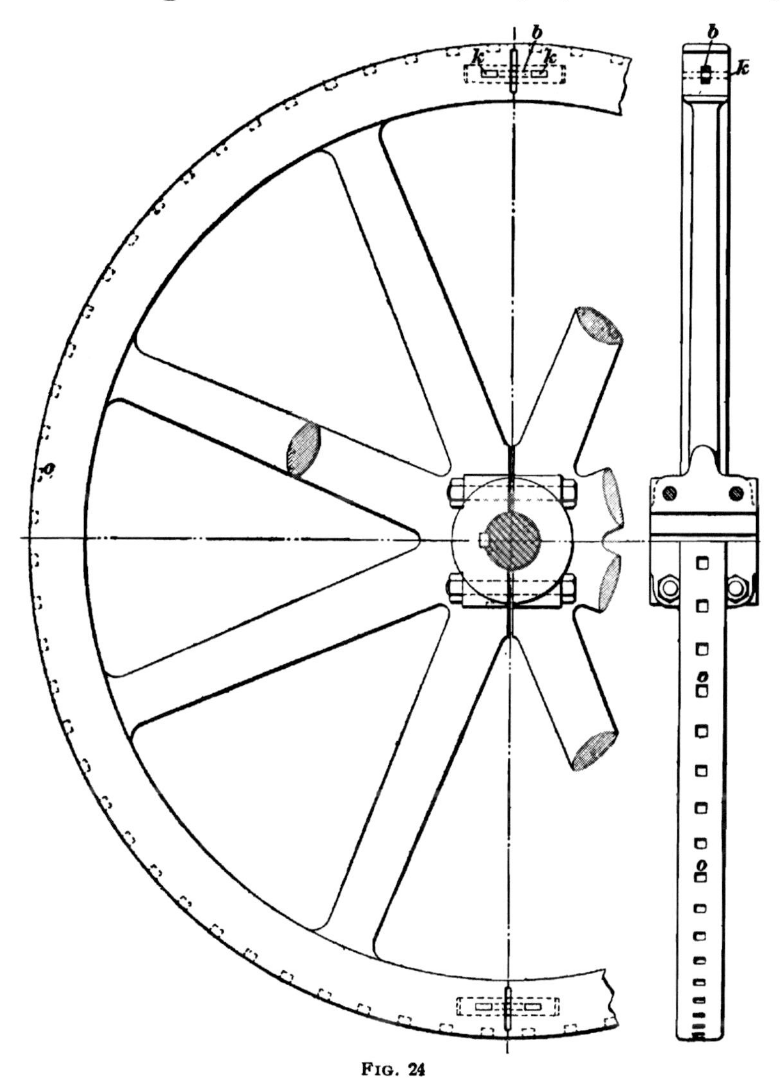

Fig. 24

the end of a bar when it is required to turn the engine, either to get the crank off the dead center in starting or for any other purpose.

ENGINE FRAMES, OR BEDS

47. The **frame,** or **bed,** of an engine is the main structure to which the other parts are attached. It is stiff and rigid, and, on account of its mass, absorbs more or less of the vibration due to the movement of the reciprocating parts. Engine beds are made in a great variety of forms, each type of engine having its peculiar type of bed.

Fig. 25 shows a substantial type of engine bed for horizontal engines. Fig. 25 (*a*) is a top view, (*b*) a horizontal section on the center line looking toward the bottom, (*c*) a front side view, (*d*) a view of the bottom, (*e*) a cross-section through the center line of the main bearing, (*f*) a cross-section on the line *O O*, (*g*) an end view, and (*h*) a cross-section through the guides on the line *P P*. The guides *L*, *L* are cast solid with the bed, and then bored out to form the bearing surface for the crosshead, which is of the form shown in Fig. 1. *I*, *I* are bosses, which form bearings for the rocker-arm shaft. These bearings are provided with brass or Babbitt bushings. The main bearing, which is separate from the frame, rests in the opening *R*.

Proportions for designing this bed are based on the diameter of the cylinder, the length of stroke, dimensions of crosshead, and length of connecting-rod, as follows:

D = diameter of cylinder.
$a = .027\,D + .1875$ inch.
$b = 1.1\,a$.
$c = 1.25\,a$.
$c' = 2.5\,a$.
$d = 1.5\,a$.
$e = 2.25\,a$.
$f = 4\,a$.
$g = 4.75\,a$, but never less than u.
$h = 6.5\,a$.
$i = 6\,a$.
$k = 12\,a$.
$k' = 13.5\,a$.
$l = 2\,a$.
m = the distance required to clear the connecting-rod.
$n = 2\,a$.
$o = .5\,a$.
$p = .75\,m$.
$q = .8\,D$ to D.
$r = 6\,a$.
$s = .5\,a$.
$t = .06\,D + .5$ inch (use the nearest standard size of bolt).
$u = 2.1\,t$

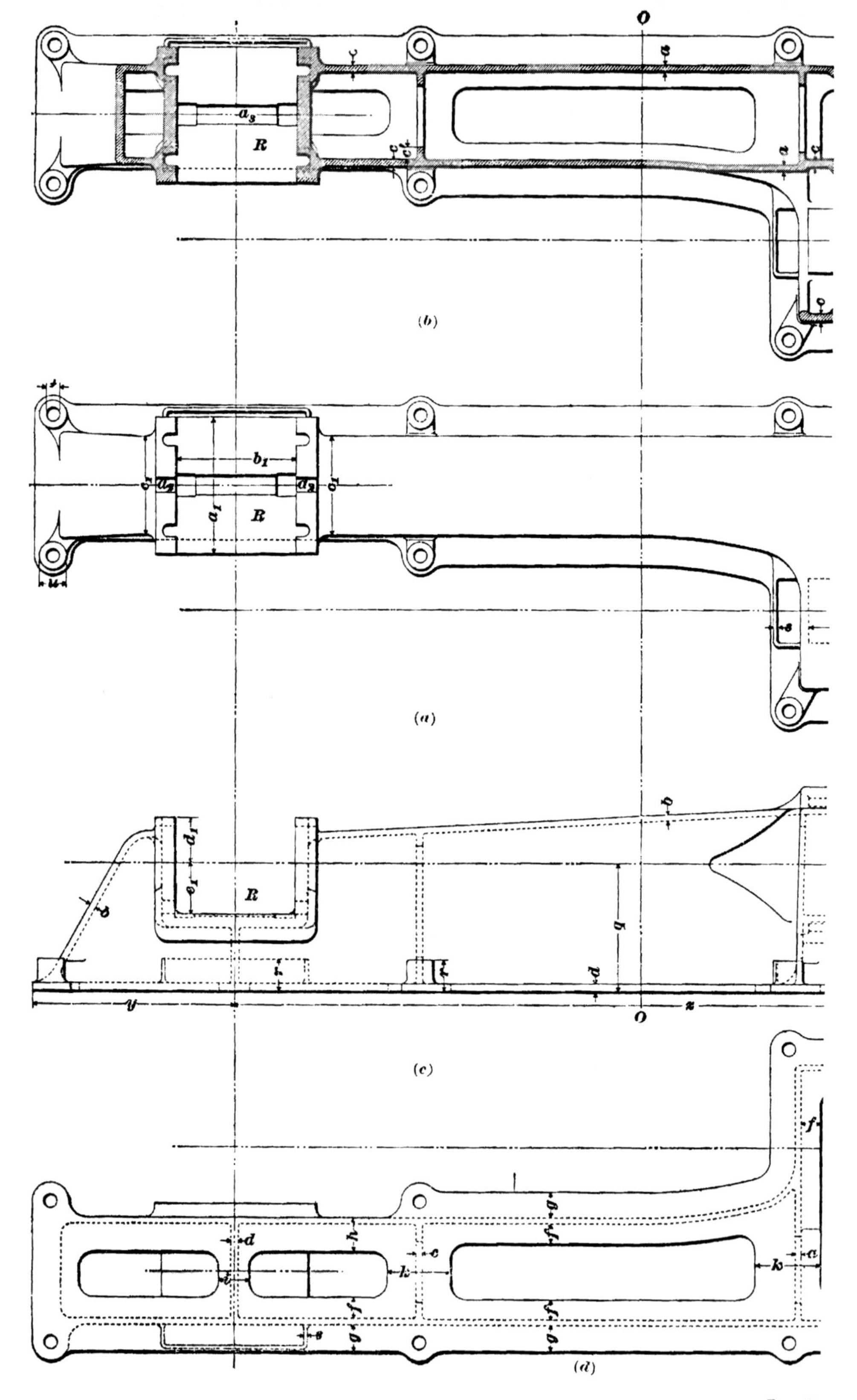

Fig. 25

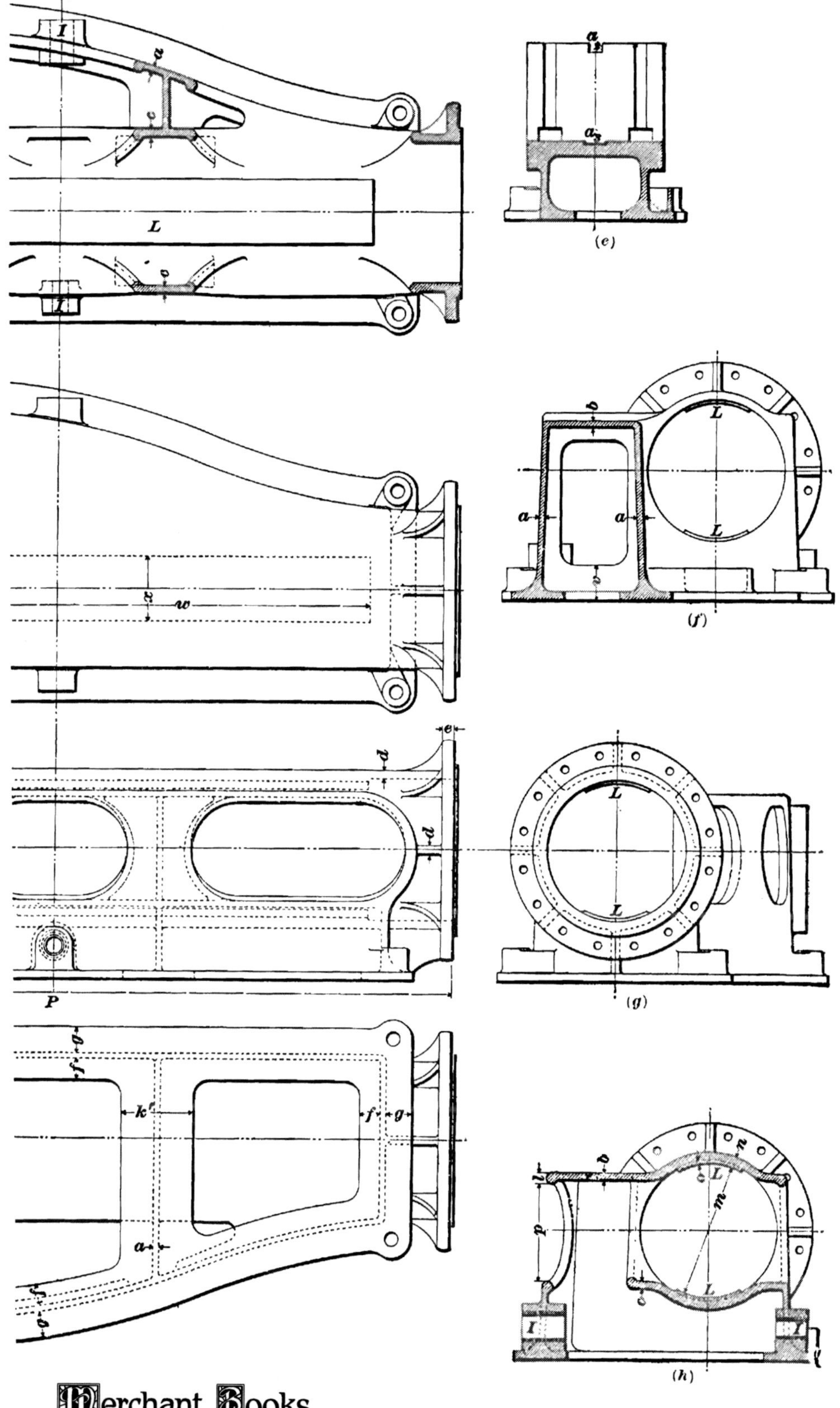
(e)
(f')
(g)
(h)

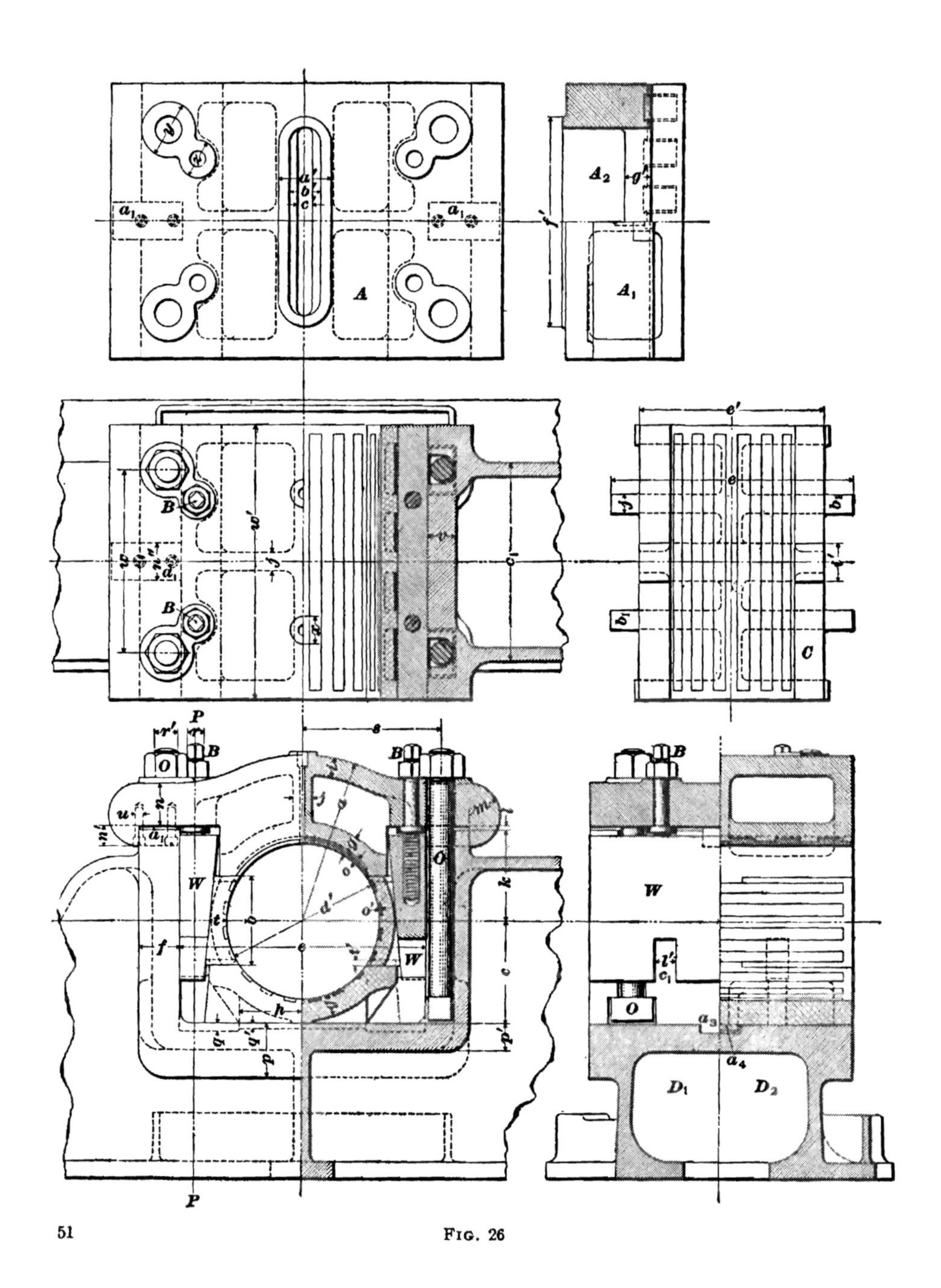

51

Fig. 26

$v = 6.5\,a$.

w = length of stroke + length of rubbing surface of crosshead − (.01 D + .1875 inch).

x = same width as crosshead.

y = about 1.3 D. In all cases, the crank must clear the bosses and nuts for the foundation bolts.

The length z must be such that the hub of the crosshead will clear the stuffingbox bolts when at the end of the stroke. Approximately, its value = length of crank + length of connecting-rod + distance from center of crosshead pin to end of crosshead hub + clearance between crosshead hub and stuffingbox bolts + the distance that the stuffingbox bolts project into the frame. This distance z is best determined by laying out the various parts to scale.

The dimensions for the seat for the main bearing are:

d' = diameter of crank-shaft journal.

$a_1 = 1.75\,d'$.

$b_1 = 1.65\,d' - .5$ inch.

$c_1 = .62\,D$.

$d_1 = .5\,d' + 1.25$ inches.

$e_1 = .66\,d'$.

The bearing for the frame shown in Fig. 25 is shown in detail in Fig. 26. Proportions for designing this bearing are:

d' = diameter of journal.

D = diameter of cylinder.

$a = d' + 1$ inch.

$a' = .2\,d' + 2$ inches.

$b = .5\,d' + 1$ inch.

$b' = .12\,d' + 1.25$ inches.

$c = .66\,d'$.

$c' = .06\,d' + .625$ inch.

$c_1 = .62\,D$.

$e = 1.65\,d' - .5$ inch.

$e' = 1.22\,d'$.

$f = .25\,d' + .375$ inch.

$f' = 1.35\,d'$.

$g = .1\,d' + .5625$ inch.

$g' = .1\,d' + 1$ inch.

$h = .85\,d'$.

$i = .1\,d' + .25$ inch.

$i' = .2\,d' + .5$ inch.

$j = .1\,d' + .25$ inch.

$k = .5\,d' + 1.25$ inches.

l = .375 inch, constant.

$l' = .1\,d' + .375$ inch.

$m = .175\,d' + .3125$ inch.

$n = .25\,d' + .25$ inch.

$n' = .1\,d' + .375$ inch.

$n'' = .2\,d' + .5$ inch.

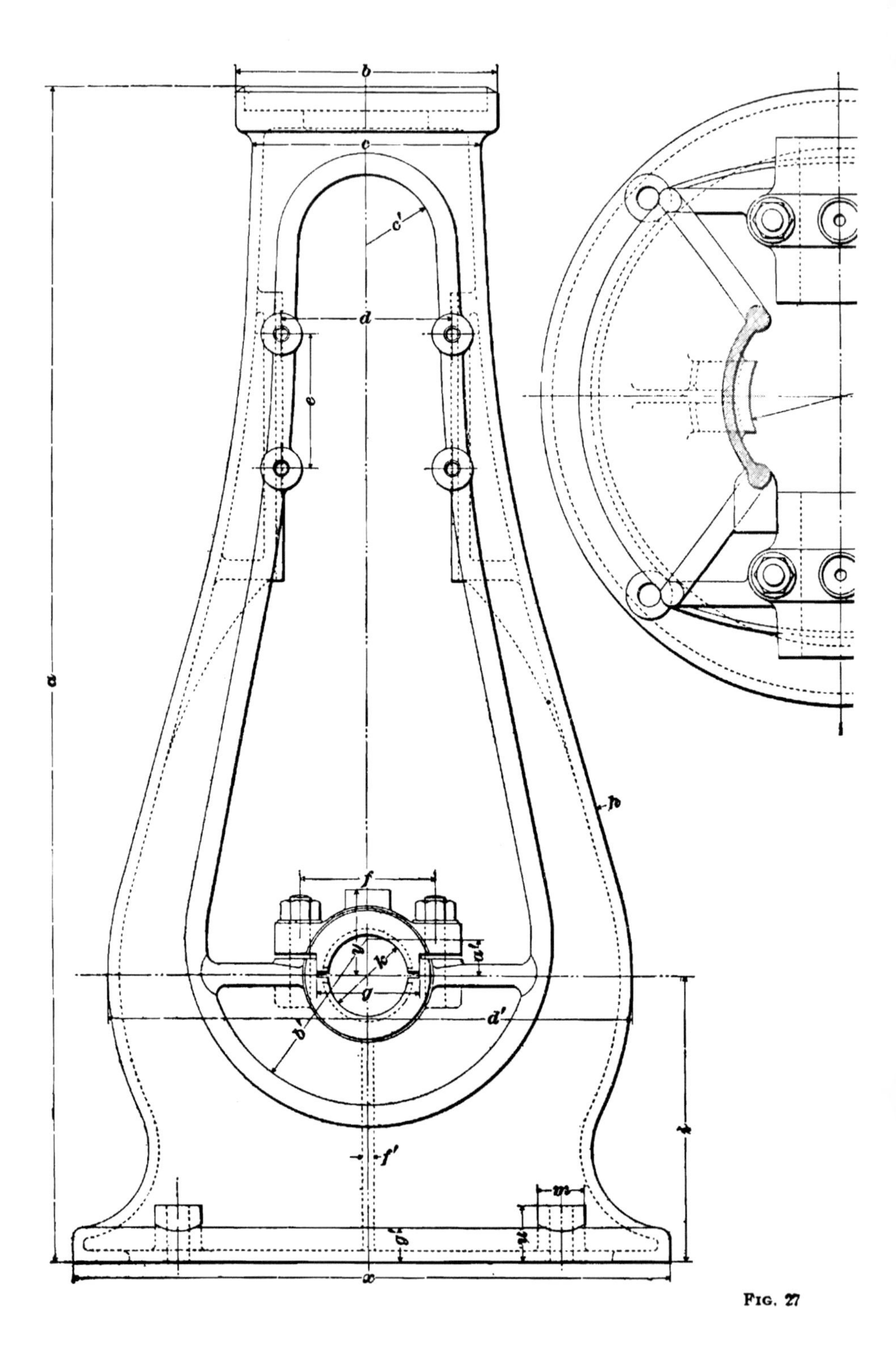

FIG. 27

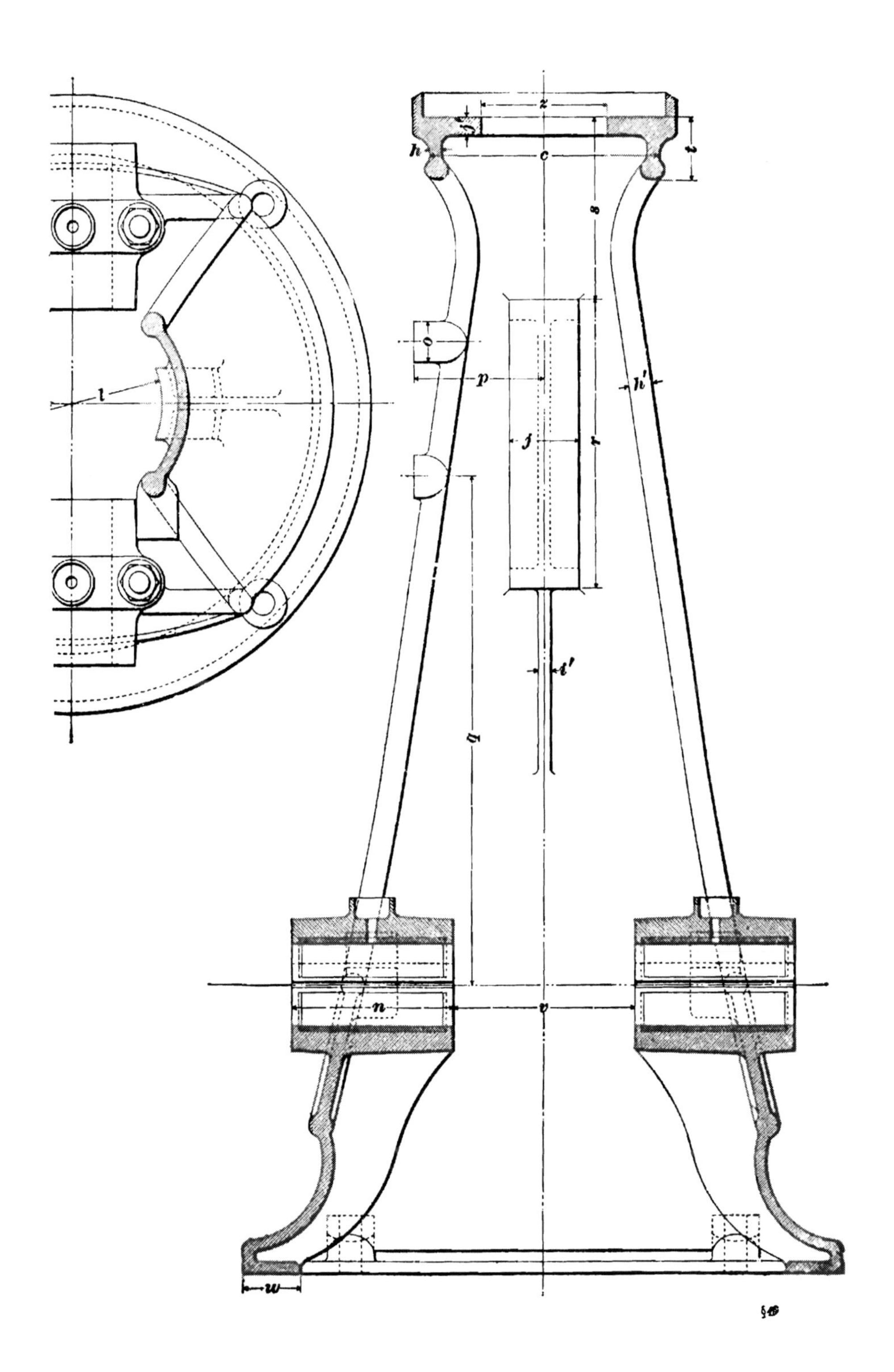

z
c
h
t
s
o
p
h'
f
r
l
d'
q
n
v
w

o = .625 inch, constant.
o' = .375 inch, constant.
p = .3 d' + .5 inch.
p' = .15 d' + .375 inch.
q = .02 d' + .5 inch.
q' = .02 d' + .25 inch.
r = .1 d'.
r' = .15 d'.
s = .9 d'.
t = .11 d'.
t' = .02 d + .25 inch.
u = .04 d' + .125 inch; use nearest standard size bolt.
v = .15 d' + .375 inch.
w = 1.2 d'.
w' = 1.75 d'.
x = 2.5 inches, constant.
y = .3 d' + .75 inch.
z = .2 d' + .5 inch.

It will be observed that the bearing shown in Fig. 26 has four seats, including the cap, which is lined with Babbitt so as to form the top seat. The two side seats can be adjusted by means of the wedges W, which are moved by the bolts B. The bearing is held to the engine bed by the **T**-head bolts O, O, which fit into slots cast in the bed for this purpose. The side and bottom seats are of brass, with Babbitt linings. A is a top view of the cap, A_1 a side view of the cap, and A_2 a section of the cap on the center line. Lugs a_1, usually of wrought iron, fit into slots a_2, Fig. 25, to prevent end motion of the cap. C, Fig. 26, is an inside, or top, view of the bottom seat. It has lugs b_1 that fit into the slots b_2 of the wedges W (see view D_1, which is a half section through the bearing on the line PP, with the wedge in place). D_2 is a half section of the bearing on the center line. The bottom seat has a lug a_4 (see section D_2) that fits into a corresponding slot in the bed. This slot is shown at a_3 in section D_1; also at a_3, Fig. 25.

48. Fig. 27 is an example of a frame for a vertical engine, as made by a well-known builder. The dimensions of this frame for various sizes of cylinders are given in Table I.

TABLE I

DIMENSIONS OF FRAME FOR VERTICAL ENGINE

Size of Engine Inches	a	b	c	d	e	f	g	h	i	j	k	l	m
3 × 5	28 1/4	6 7/8	6	3 3/4	3 1/4	3 5/8	2 1/4	1/4	6 3/4	1 1/4	1 3/4	4 1/2	1 1/2
4 × 6	34 3/4	7 7/8	7 1/4	5 1/4	4 1/2	4 1/2	3	3/8	7 3/4	1 3/4	2	5	1 1/2
5 1/2 × 7	40 1/4	9 1/2	8	6	5 1/2	5	3 5/8	3/8	10	2 1/2	2 1/2	6	1 3/4
7 × 9	50 1/4	11 1/2	10	7 1/2	5 3/4	6 1/4	4	1/2	12 3/8	2 1/2	2 15/16	7 7/8	2
9 × 12	61 1/4	14 1/4	12 1/2	9	7 1/4	9	6 3/4	5/8	13 3/4	3 1/2	3 15/16	9 1/2	2

Size of Engine Inches	n	o	p	q	r	s	t	u	v	w	x	y	z
3 × 5	3 3/8	1	3 1/16	12 3/8	7 3/4	2 15/16	1 3/4	1 5/8	3 3/8	1/2	12 3/4	2	3 5/8
4 × 6	4 7/16	1 1/2	3 7/8	13 3/4	9 3/8	4 3/4	2 5/8	1 5/8	4 1/8	5/8	15	2 1/4	4 1/2
5 1/2 × 7	5 13/16	1 1/2	4 3/8	15 1/2	10 1/2	4 3/4	2 1/4	2 5/8	4 7/8	1	18	3 1/2	4 3/4
7 × 9	5 5/8	1 3/4	5	21 5/8	12 1/4	7 3/4	3 3/4	3	6 3/4	2 1/2	26 1/4	4 1/4	5 1/2
9 × 12	5 1/2	2	5 3/4	30 1/8	17 1/2	7 1/2	4 3/4	3 1/2	11	3 1/4	26 1/2	6	7 1/2

Size of Engine Inches	a′	b′	c′	d′	e′	f′	g′	h′	i′	j′
3 × 5	11/16	4	2 3/16	11 1/2	5/8	3/8	1	1/2	3/8	1/2
4 × 6	15/16	4 5/8	2 1/2	13 1/2	5/8	1/2	1 1/8	3/4	1/2	3/4
5 1/2 × 7	1	5 1/4	2 5/8	15 3/4	3/4	1/2	1 1/4	3/4	1/2	3/4
7 × 9	2 3/16	7 7/8	3 11/16	23	7/8	9/16	1 1/2	1	9/16	1
9 × 12	3/4	7 5/8	4 7/16	23 1/2	7/8	5/8	1 3/4	1 1/4	5/8	1 1/4

e′ = size of foundation bolt. For 9″ × 12″ engine, a disk center crank is used.

EXAMPLES OF ENGINE PROPORTIONS

49. It has been previously stated that when a standard line of engines is to be manufactured, the rules and formulas

TABLE II

PROPORTIONS OF CORLISS ENGINES

Diameter of Cylinder Inches	Stroke Inches	Revolutions per Minute	Piston Speed Feet per Minute	Diameter of Shaft Journal Inches
10	24	90	360	5
12	24	90	360	$5\frac{1}{2}$
14	30	80	400	7
16	30	80	400	$7\frac{1}{2}$
16	36	70	420	8
18	36	70	420	$8\frac{1}{2}$
20	36	70	420	9
16	42	65	455	$8\frac{1}{4}$
18	42	65	455	$8\frac{3}{4}$
20	42	65	455	$9\frac{1}{4}$
20	48	60	480	$9\frac{1}{2}$
22	48	60	480	10
24	48	60	480	10
22	54	60	540	$10\frac{1}{4}$
24	54	60	540	$10\frac{3}{4}$
26	54	60	540	$11\frac{1}{4}$
24	60	60	600	11
26	60	60	600	$11\frac{1}{2}$
28	60	60	600	12
30	60	60	600	$12\frac{1}{2}$
32	60	60	600	13
34	60	60	600	$13\frac{1}{2}$
36	60	60	600	14

for the design of the various parts need not be applied to each individual engine. It is found that under the conditions in which the engines are to work, a certain ratio may be assumed to exist between the sizes of the parts. For example, a certain line of engines work uniformly at a steam pressure of 75 pounds, and the length of the piston rod bears a fixed relation to the length of stroke. Under these circumstances, the diameter of the piston rod may be a fixed fraction of the diameter of the cylinder for all sizes, and it is

TABLE III

PROPORTIONS OF CORLISS ENGINES

Diameter of Cylinder	Diam. and Length of Crank-pin	Diameter and Length of Cross-head Pin	Diameter of Valve Stem	Depth of Piston	Diameter of Piston Rod	Width of Crank-Disk	Clearance of Piston
Inches	Inches	Inches	Inches	Inches	Inches	Inches	Inch
10	$2\frac{1}{2}$	$1\frac{5}{8} \times 2\frac{1}{2}$	$\frac{7}{8}$	4	$1\frac{5}{8}$	$2\frac{1}{2}$	$\frac{1}{4}$
12	3	2×3	1	$4\frac{1}{2}$	2	3	$\frac{1}{4}$
14	$3\frac{1}{2}$	$2\frac{3}{8} \times 3\frac{1}{2}$	$1\frac{1}{8}$	5	$2\frac{1}{4}$	$3\frac{1}{2}$	$\frac{1}{4}$
16	4	$2\frac{5}{8} \times 4$	$1\frac{1}{4}$	$5\frac{1}{2}$	$2\frac{1}{2}$	4	$\frac{1}{4}$
18	$4\frac{1}{2}$	$3 \times 4\frac{1}{2}$	$1\frac{3}{8}$	6	$2\frac{7}{8}$	$4\frac{1}{2}$	$\frac{1}{4}$
20	5	$3\frac{3}{8} \times 5$	$1\frac{1}{2}$	$6\frac{1}{2}$	$3\frac{1}{4}$	5	$\frac{1}{4}$
22	$5\frac{1}{2}$	$3\frac{5}{8} \times 5\frac{1}{2}$	$1\frac{1}{2}$	7	$3\frac{1}{2}$	$5\frac{1}{2}$	$\frac{5}{16}$
24	6	4×6	$1\frac{5}{8}$	$7\frac{1}{2}$	$3\frac{7}{8}$	6	$\frac{5}{16}$
26	$6\frac{1}{2}$	$4\frac{1}{4} \times 6\frac{1}{2}$	$1\frac{3}{4}$	$7\frac{3}{4}$	$4\frac{1}{8}$	$6\frac{1}{2}$	$\frac{5}{16}$
28	7	$4\frac{5}{8} \times 7$	$1\frac{3}{4}$	$8\frac{1}{4}$	$4\frac{1}{2}$	7	$\frac{5}{16}$
30	$7\frac{1}{2}$	$5 \times 7\frac{1}{2}$	$1\frac{7}{8}$	$8\frac{1}{2}$	$4\frac{7}{8}$	$7\frac{1}{2}$	$\frac{5}{16}$
32	8	$5\frac{1}{4} \times 8$	$1\frac{7}{8}$	$8\frac{1}{2}$	$5\frac{1}{8}$	8	$\frac{5}{16}$
34	$8\frac{1}{2}$	$5\frac{5}{8} \times 8\frac{1}{2}$	2	$8\frac{3}{4}$	$5\frac{1}{2}$	$8\frac{1}{2}$	$\frac{3}{8}$
36	9	$5\frac{7}{8} \times 9$	2	9	$5\frac{3}{4}$	9	$\frac{3}{8}$

only necessary, therefore, to multiply the cylinder diameter by this fraction to find the diameter of the piston rod.

Tables II to V, inclusive, give the proportions of a standard line of Corliss engines made by a leading manufacturer. They will serve to illustrate the use of fixed proportions in designing, and will also furnish valuable examples of good, modern practice.

TABLE IV

PROPORTIONS OF CORLISS ENGINES

Diameter of Cylinder	Steam Inlet Area	Exhaust Area	Diameter of Valve	Inlet Ports	Exhaust Ports	Width of Steam Chest	Width of Exhaust Chest	Depth of Steam and Exhaust Chests	Diameter of Steam Pipe	Diameter of Exhaust Pipe
Inches	Square Inches	Square Inches	Inches	Inches	Inches	Inches	Inches	Inches	Inches	Inches
10	4.71	7.85	$3\frac{1}{8}$	$\frac{1}{2} \times 10$	$\frac{13}{16} \times 10$	5	$6\frac{1}{2}$	$1\frac{3}{4}$	$2\frac{1}{2}$	$3\frac{1}{2}$
12	6.79	11.31	$3\frac{1}{2}$	$\frac{9}{16} \times 12$	$\frac{15}{16} \times 12$	6	$7\frac{1}{2}$	2	3	4
14	9.23	15.39	$3\frac{7}{8}$	$\frac{11}{16} \times 14$	$1\frac{1}{8} \times 14$	7	9	$2\frac{1}{4}$	$3\frac{1}{2}$	$4\frac{1}{2}$
16	12.00	20.00	$4\frac{1}{4}$	$\frac{3}{4} \times 16$	$1\frac{1}{4} \times 16$	8	10	$2\frac{1}{2}$	4	5
18	15.27	25.44	$4\frac{5}{8}$	$\frac{7}{8} \times 17$	$1\frac{1}{2} \times 17$	9	$11\frac{1}{2}$	$2\frac{3}{4}$	$4\frac{1}{2}$	6
20	18.85	31.42	5	1×19	$1\frac{11}{16} \times 19$	10	$12\frac{1}{2}$	3	5	$6\frac{1}{2}$
22	22.80	38.00	$5\frac{3}{8}$	$1\frac{1}{8} \times 20\frac{7}{8}$	$1\frac{7}{8} \times 20\frac{7}{8}$	11	$14\frac{1}{2}$	$3\frac{1}{4}$	$5\frac{1}{2}$	7
24	27.14	45.24	$5\frac{3}{4}$	$1\frac{3}{16} \times 22\frac{7}{8}$	$2 \times 22\frac{7}{8}$	12	15	$3\frac{1}{2}$	6	$7\frac{1}{2}$
26	31.85	53.09	$6\frac{1}{8}$	$1\frac{5}{16} \times 24\frac{3}{4}$	$2\frac{1}{8} \times 24\frac{3}{4}$	13	$16\frac{1}{2}$	$3\frac{3}{4}$	$6\frac{1}{2}$	$8\frac{1}{2}$
28	36.95	61.58	$6\frac{1}{2}$	$1\frac{7}{16} \times 25\frac{1}{2}$	$2\frac{1}{2} \times 25\frac{1}{2}$	14	$17\frac{1}{2}$	4	7	9
30	42.41	70.69	7	$1\frac{9}{16} \times 27\frac{1}{2}$	$2\frac{5}{8} \times 27\frac{1}{2}$	15	19	$4\frac{1}{2}$	$7\frac{1}{2}$	$9\frac{1}{2}$
32	48.26	80.43	$7\frac{1}{2}$	$1\frac{5}{8} \times 29\frac{1}{4}$	$2\frac{3}{4} \times 29\frac{1}{4}$	16	20	5	8	$10\frac{1}{2}$
34	54.48	90.79	8	$1\frac{3}{4} \times 31\frac{1}{4}$	$2\frac{7}{8} \times 31\frac{1}{4}$	17	$21\frac{1}{2}$	$5\frac{1}{2}$	$8\frac{1}{2}$	11
36	61.07	101.79	$8\frac{1}{2}$	$1\frac{7}{8} \times 33$	$3\frac{1}{8} \times 33$	18	$22\frac{1}{2}$	6	9	$11\frac{1}{2}$

TABLE V

PROPORTIONS OF CORLISS ENGINES

Diameter of Cylinder Inches	Thickness of Cylinder Inches	Thickness of Chests Inches	Thickness of Valve Chamber Inches	Bearing of Valves Inches
10	$\frac{3}{4}$	$\frac{5}{8}$	$\frac{3}{4}$	$\frac{7}{8}$
12	$\frac{13}{16}$	$\frac{5}{8}$	$\frac{3}{4}$	$1\frac{1}{16}$
14	$\frac{7}{8}$	$\frac{11}{16}$	$\frac{13}{16}$	$1\frac{1}{8}$
16	$\frac{15}{16}$	$\frac{3}{4}$	$\frac{7}{8}$	$1\frac{1}{4}$
18	1	$\frac{13}{16}$	1	$1\frac{3}{8}$
20	$1\frac{1}{16}$	$\frac{7}{8}$	$1\frac{1}{16}$	$1\frac{1}{2}$
22	$1\frac{1}{8}$	$\frac{15}{16}$	$1\frac{1}{8}$	$1\frac{5}{8}$
24	$1\frac{3}{16}$	$\frac{15}{16}$	$1\frac{1}{8}$	$1\frac{3}{4}$
26	$1\frac{1}{4}$	1	$1\frac{3}{16}$	$1\frac{7}{8}$
28	$1\frac{1}{4}$	1	$1\frac{1}{4}$	2
30	$1\frac{5}{16}$	$1\frac{1}{16}$	$1\frac{5}{16}$	$2\frac{1}{8}$
32	$1\frac{3}{8}$	$1\frac{1}{8}$	$1\frac{3}{8}$	$2\frac{1}{4}$
34	$1\frac{3}{8}$	$1\frac{1}{8}$	$1\frac{3}{8}$	$2\frac{3}{8}$
36	$1\frac{3}{8}$	$1\frac{1}{8}$	$1\frac{3}{8}$	$2\frac{1}{2}$

50. An inspection of Tables II to V, inclusive, shows that the following rules are used in designing the line of engines considered:

Let D = diameter of cylinder. Then,

$$\text{diameter of shaft} = .34\,D + 2\tfrac{1}{2} \text{ inches, nearly}$$

A more exact but less simple formula, where L is the stroke, is

$$\text{diameter of shaft} = .26\sqrt[3]{D\,L} + 2\tfrac{1}{2} \text{ inches}$$

$$\text{Diameter of crankpin} = \frac{D}{4}.$$

$$\text{Length of crankpin} = \frac{D}{4}.$$

$$\text{Length of crosshead pin} = \frac{D}{4}.$$

Diameter of crosshead pin = diameter of crankpin × .65.

Diameter of steel valve stem = $.19\sqrt[3]{D^2}$.

Depth of piston

$= \frac{D}{4} + 1\frac{1}{2}$ inches for D less than 24 inches;

$= \frac{D}{8} + 4\frac{1}{2}$ inches for D greater than 24 inches.

Diameter of piston rod $= .16\ D$.

Width of crank-disk $= \frac{D}{4}$.

Area of steam port $= .06 \times$ area of cylinder.

Area of exhaust port $= .1 \times$ area of cylinder.

Diameter of valve $= \frac{3\ D}{16} + 1\frac{1}{4}$ inches.

Width of steam chest $= \frac{D}{2}$.

Width of exhaust chest $= .63\ D$.

Depth of steam and exhaust chests

$= \frac{D}{8} + \frac{1}{2}$ inch for D less than 28 inches;

$= \frac{D}{4} - 3$ inches for D greater than 28 inches.

Diameter of steam pipe $= \frac{D}{4}$.

Diameter of exhaust pipe $= .31\ D$.

Thickness of cylinder $= .028\ D + \frac{1}{2}$ inch.

Thickness of chests $=$ thickness of cylinder $\times .8$.

Thickness of valve chamber

$=$ thickness of chest $+ \frac{1}{8}$ inch for $D = 10$ to 16 inches;

$=$ thickness of chest $+ \frac{3}{16}$ inch for $D = 18$ to 26 inches;

$=$ thickness of chest $+ \frac{1}{4}$ inch for $D = 28$ to 36 inches.

Bearing of valve $=$ diameter of valve $\times .3$.

CPSIA information can be obtained at www.ICGtesting.com
Printed in the USA
BVOW060041301111

277192BV00003B/84/P